Every Day For
My Daughter

This book is dedicated to all my family and friends. I hope it touches your heart as you have mine.

Every Day For My Daughter

Faith . . . in the face of a deadly flesh-eating infection.

Timothy N. Cole

authorHOUSE®

AuthorHouse™
1663 Liberty Drive
Bloomington, IN 47403
www.authorhouse.com
Phone: 1-800-839-8640

All names have been changed to protect the innocent
Scripture quotations are taken from the King James Version (KJV) of the Bible-Public Domain

Published by AuthorHouse   01/11/2013

ISBN: 978-1-4817-0144-0 (sc)
ISBN: 978-1-4817-0143-3 (hc)
ISBN: 978-1-4817-0142-6 (e)

Library of Congress Control Number: 2012924160

I would like to thank Dr. Katie for her courtesy and candor during our interviews. And a special 'thank you' to Scott Messinger for the CCR line in the final chapter; a small tribute to our creative sessions.

Williamsport, Pennsylvania

Thy sun shall no more go down, neither shall thy moon withdraw itself: for the Lord shall be thine everlasting light, and the days of thy mourning shall be ended.

-Isaiah 60:20

February 22, 1990

Keith Bryant Cole was broken. He was as broken as any man has ever been. And now in the dark solitude of his bedroom, the difference between emotional suffering and physical pain, grief and guilt, an exhausted man and an idle father, were no longer discernable to him. His psyche was shattered until the cracks nearly lacerated his flesh. The only tears that remained in his body before the onset of evening had already trickled down his cheeks, leaving his eyes stingingly dry and tired. His belief in God was now spawned by desperation as he resisted the notion that the life-force he knew as Samantha no longer existed in any form. Heaven had to exist so that Samantha, the tiny girl his ex-wife nicknamed Cricket, had a place to go. That alone was the crux of his faith.

He cued a CD to *Vincent,* a song by Don McLean about the artist, Vincent Van Gogh, and quietly placed a chair next to the open door of his room. *His room?* If his heart wasn't so broken, the thought would be laughable. His room. Nothing was his anymore. Not his house, not his daughter, not his sense of security (no matter how frail its initial construction) and most assuredly not the drain-vortex he called his life.

The dirgeful glow of a three quarter moon coated the plaster walls in a portentous pale. It twisted the shadows of all forms of objects into unshaped bodies, bowed in remorse.

His parents were asleep in their bedroom on the far side of the house, stirring them would bring an end to his plan and more attention than he cared to have bestowed upon himself. He was appreciative of their hospitality for allowing their thirty-year-old son to live with them again, a situation that did little for his sense of pride. His marriage to Lori had crumbled, and he had nowhere to go but back home to his parents. And then came the unspeakable; a thunder-strike that would shake the foundation of the rest of his life. Samantha, the paragon of childhood innocence, had lost her life in a day that lasted a horrific lifetime. And since her death, everyone appeared uncomfortable in Keith's presence. They desperately tried to say the right thing, or made a deliberate attempt not to say anything, as to avoid making a verbal faux pas. Not that saying anything positive or otherwise could ever possibly bring solace to his situation. Words spoken were designed to stroke some unsuspecting neuron into releasing a shallow-buried, positive memory or life-lesson that could rise up and smite the dark lord of depression. But words became meaningless; Keith's brain had been rewired. All circuits were now redirected to Samantha. The moon was no longer Earth's nearest celestial neighbor; it was the object of Samantha's fascination. It was a bright point of focus in the night sky as seen through a father and daughter's bedroom window.

The sounds of *Vincent* provided a bubbling-brook dampening to the anguished screams of the civil war battle that raged between the hemispheres of Keith's mind. The gentle, melodic vocals that had once echoed in an empty 1970's recording studio now eased from small speakers; the emotion from the singers voice transcending nearly twenty years to touch Keith's life in a way that was never intended. The softly plucked strings vibrated in the background as an occasional lyric brushed aside the rampant thoughts of Keith's mourning.

"Starry, starry night. Paint your palette blue and grey . . ."

He tied the ends of two neckties together in a double knot, creating a fashionable rope. The end that was an indistinguishably bland gray color with no design, one that he no longer wore, he secured to the doorknob. He created a slipknot on the end that boasted a blue, diamond pattern and stepped up onto the chair, pulling the loop down over his head. He wanted the pain to stop, pain that his tears could never wash away.

". . . catch the breeze and the winter chills . . . in colors on the snowy linen land."

The lines of the song bit at his memory with icy insinuation. His two-year-old daughter had drowned in an in-ground swimming pool—in water made frigid by the wintery breath of a January in Pennsylvania. A swimming pool that once held fond memories of family gatherings and frolicking summer parties now reminded him of the cold nature of life.

Now it's silent. Everything is silent. The wind that rustled the grass. The splashing of the water. The giggles of my little girl. The frost made certain that everything was dead. And I wasn't there to stop it. I let it happen. I let her down. I let her drown. I should have been there to save her.

His imagination had played the scenario so often, the one that denied reality—the scenario which provided him the oblivious satisfaction that Samantha was still walking hand-in-hand with him through life, until he began dreaming it. In these dreams, the child herself reassured him that she was okay. His brain attempting to fulfill a wish of which, it could not bear to let go. But history had taught him of reality's cruelty and God's inexplicable apathy toward intervention. And now, He had placed Keith at a crossroads where either path led to suffering. Live with the pain of his loss and pray he will be reunited with Samantha on the other side, or commit the sin of suicide and risk never seeing his daughter again. No answers were granted him; no signs of divine guidance; no heavenly host, quadric-winged seraphs delivering the message, "Don't lose faith, for God is with thee." His heart was so gashed, he would have accepted the guidance of any supreme deity, by any name; God, Allah, Yahweh . . . He would have sought healing through the teachings of the Christ, the nothingness of Zen meditation, self-preservation through Dianetics and Scientology, simply the power of positive thinking or whatever lights man's way out of the darkness of misery. However, his pain had isolated him.

". . . *how you suffered for your sanity . . .*"

Samantha's accident happened on a Sunday following one of his scheduled weekends with his three children. Normally these visitations were cherished, but on that day, he was exhausted from long hours at his job working for a chain of franchised fast-food restaurants. He was supervising several locations, and while the pay was good, the hours were grueling. He was near nodding off for most of the afternoon and Samantha and her two brothers, Jesse and Shilo were becoming restless from boredom. He compelled himself to take the children to visit the dogs at an animal shelter close to his estranged wife's home. While advancing her studies of the simplicity of canine existence, Samantha stepped in dog crap—a sign to Keith that the day was not going to get any better. The atmosphere in the car bristled with the usual banter from his two sons. Samantha rode in silence. When they arrived at the ranch-style home where he had once lived with Lori and their kids, Samantha was distraught as he reached in the car to release her seatbelt. She was crying, worried that she had done something wrong to make her daddy angry. He tried to reassure her that he wasn't angry over her misstep outside the kennel, but the child requested her own reassurance by smacking her lips; a signal for Daddy to kiss her.

"... *weathered faces lined in pain* . . ."

The water was so cold and you were so small. I should have been there. How can I ever live with this? I just want the pain to go away. Please, let me have peace.

Later that evening he was summoned to the emergency room. His daughter had been pulled from the swimming pool that was once the focal point of his and Lori's home. The prognosis was not good. Samantha was unconscious and on a respirator with increased swelling around the brain. A 'Life Flight' helicopter transferred her to the Geisinger Medical Center and Children's Hospital where she would lose her battle to survive the following day. Samantha's life was a brief two years, but her impression on the life fabric of those who knew her and those who would one day hear her story, would reach far beyond anything her daddy could imagine in the desperate moment in which he was locked.

And as the music swirled within his dizzied thoughts, Keith heard the lyrics for which he had waited.

"... and when no hope was left in sight, on that starry, starry night ... you took your life as lovers often do ..."

Keith eased himself down from the chair and the tie around his neck tightened to a painful grip. His carotid arteries pinched and his head thumped in agony.

Oh, God, this hurts. Let me die quickly. Let it be over.

The fabric of the tie imposed its strength into the flesh of his neck with slicing indifference.

"... this world was never meant for one as beautiful as you."

Is this wrong? Is this really what I want to leave behind? What about my boys?

"... frameless heads on nameless walls ... with eyes that watch the world and can't forget."

My sons need me. How can I let them down like this?

His eyes, though blurry from tears, searched the room that he had once danced around with his precious little girl in his arms. On those occasions, they would play *Sam,* a song by Olivia Newton-John. He would hold her close and sing in her ear—father soothing daughter; daughter soothing father. Now, in the wake of Samantha's passing, moments that were once fond memories became the most painful.

If I do this, I may never see Samantha again. I can't go through with it.

He reached up, gripped the top of the door with both hands and lifted his weight enough to ease the tension. He searched with his feet until he found the chair and regained his perch. He grabbed at the tie and in frantic frustration, undid the knot around his neck then cast it aside.

Damn you. Give me something. Give me a sign that she's in a better place. I want to believe. I need to believe.

The contemplation of suicide was behind him, but the pain of his loss and his journey of faith was only beginning. He would be broken even further before it was done.

". . . perhaps they never will."

November 25, 2003

Restless legs and pale green scrubs are always sure signs of an expectant father. Keith sat in a blue plastic chair outside of the operating room as his second wife, Stacey, was prepped for the delivery—his knees pumping up and down like the divine mechanized synchronicity of pistons in a Mercedes. Thirteen years had passed and the changes in his life took him far from the disjointed days following Samantha's death. His two boys were now teenagers; he had remarried and settled, comfortably, into an upper-middle class existence. He and Stacey owned a home in a rural fringe of the city, sharing their neighborhood with the likes of doctors and lawyers. But despite every reason to be jubilant, the intersecting lines of life were eliciting emotions of pessimism from within him. He recalled his conversation with Stacey as they ate dinner the previous evening.

At a small table in the middle of a dimly lit restaurant, the atmosphere surrounding them had chattered with voices, clattered with plates and reverberated the serene sounds of Bach. Stacey's deep brown eyes sparkled with new life and her creamy skin cast a subtle glow, but Keith was too distracted to notice. The journey to fathering his second daughter had become an unfortunate reminder of the joy and tragedy that was the life of Samantha.

Stacey wanted a girl, but the idea of raising a daughter made Keith feel as though he would somehow betray Samantha's memory. He kept his thoughts to himself, observing quietly as Stacey decorated the nursery in the neutral theme of Noah's Ark before she slowly began adding girls' clothing to the closet. When sonography revealed the sex of the infant to be female, he was happy for his wife, burying the pain of Samantha so as not to spoil their moment. Sonography had also revealed the daughter was in a 'feet first' position. The baby would be breech if natural birth was attempted, so a Caesarean section was scheduled on the Monday before Thanksgiving resulting in further thoughts of Samantha, as she was also born by C-section.

"Why are you so glum?" Stacey had asked with an appeasing smile. "You should be happy."

Keith countered with a rueful smile. "Stacey, you have no idea what you're in for."

She pressed her hand against her tummy. "What is your problem?" she spat. "I don't need to hear this on the day before I deliver."

"Being a parent is the worst and the best of life all rolled into one."

"I know it's not going to be easy," her tone emanated like the pout of a willful child. "We're going to have a baby and it's a little late to be back-stepping now." She suddenly fell silent, a signal to Keith that she was unhappy with the direction of the conversation.

Signal received. "Are you sure you don't want to name her Peyton?" he asked. "It's not too late to change your mind."

She passed off his request with a sardonic shake of her head. "We're not naming our daughter after your football."

"Peyton Manning plays football for the Colts—he's not actually a football himself."

At the time of the announcement, Keith had suppressed his feelings of Samantha by focusing on the positive aspects of child rearing, such as what to name his unborn girl. He then made a proposal that touched off a brief onomastics debate. Keith did what any self-respecting male would do in light of a forthcoming daughter—he made an attempt to name her after a player from his favorite football team. Stacey sacked the idea.

"You know what I mean. We're not naming her Peyton or Grace, its Isabella."

Grace was Keith's second choice before Stacey suggested the name Isabella. And so, knowing that compromise was the first lesson any man learns about marriage, the couple reached an agreement. The woman was always right.

"I'm the one who had to gain all the weight and throw up every day for the first three months, so I name the baby."

Keith fell silent. He couldn't defend his feelings nor explain them; and he hadn't the strength to smile and pretend all was well. The death of a child cannot be explained nor imagined, in order to understand the black hole that forever abides in the pit of your stomach; one would have to experience it. The very idea ate at the lining of his stomach until nausea locked his jaw before every bite of his food. The fear that it may happen again, the rerun of a nightmare which haunted him to this day, seemed far too real. It was the Zuni demon doll that had come to life to relentlessly stab Karen Black with his spear in the 1975 movie *Trilogy of Terror*.

Only now, it was not a 'made for TV movie' disturbing him during his childhood. The real-life Keith had replaced actress Karen Black as the target for its torment and the spear yielding demonic doll was a melding of guilt, remorse and fear. He imagined a curse, perhaps cast upon him for previous unkind acts. Or even a long-standing punishment struck down from the heavens by God himself. No matter the nature of its origin, he did not wish to expose Stacey nor their unborn child as a possible target.

Now outside the operating room, Keith rubbed his palms together and stared down at the floor longing to find something to focus on, other than the impending C-section. The squeaking of tennis shoes drew his attention to a familiar face. Dr. O'Hara, who was to be Isabella's pediatrician, had set aside time to attend the blessed event. She stopped in front of Keith with a reassuring smile and a pat on the shoulder.

"Are you nervous?" She asked.

Keith nodded. "I am for Stacey. I don't want to see her in pain."

"I'll check on her for you."

"Thank you."

Dr. O'Hara disappeared through the door and Keith's focus returned to the gray flecks in the tiled floor. For a moment, his legs became still as he took a deep breath and then released it in a quick burst. The contingencies of the C-section were knotting his stomach. He could remember the birth of Samantha. During the procedure, his first wife, Lori, cried out because she could feel the surgeon's knife as it sliced through her abdomen. The anesthesiologist made an adjustment to Lori's IV and the discomfort subsided. Lori was a woman of iron resolve with a high threshold for pain. If it brought her to tears, she must have been in anguish.

To Keith, the thought of seeing Stacey suffer a similar ordeal was unbearable.

Dr. O'Hara poked a masked face back into the hallway. "They're ready for you!" she said cheerfully.

Attempting to balance on shaky legs, Keith hurriedly made his way toward the operating room. Beyond the door was an event he both longed for and dreaded. *Please let it go quickly*, he thought.

Stacey lay on a sheet which she was quite certain had spent some time in a liquid nitrogen tank. Another sheet attached to a portable metal rack hung down lazily on her chest, strategically aligned to protect her from seeing more than she would care to remember of her child's birth.

Giving birth to a child; becoming a mother for the first time; undergoing surgery while you are awake—each situation would be cause for anxiety. She was about to face all three as she attempted to focus less on what her nurses training had taught her and more on quelling the fluttering in her stomach. Isabella was occasionally active, but what she was experiencing was not the motion of a fetus. Her stomach was set on intermittent vibrate and goose bumps dotted her skin. *I'll be the first Thanksgiving turkey to be carved.*

The questions surrounding the days, months and years to follow were a dense forest of self-doubt. To Stacey, motherhood was a road fraught with land-mines and possible miscues. Love and preparation were all she had—she would have to gain the experience as she went. However, it was her lack of experience that concerned her most.

Keith was of little help in consoling her. He seemed to be battling his own personal demons; ones that she had no intentions of entertaining. Early in the morning he had made a weak attempt at reassuring her.

She had stepped from the shower and wrapped a thick, warm towel around her, pulling it tight around her shoulders in an attempt to shelter herself from her fears. She eased out a sigh as she gazed at her reflection in the mirror. Her belly popped out through the towel like a billboard standing out among the trees on the edge of a roadway—announcing the arrival of a new product. She yearned to be reunited with the flat tummy that was once a source of pride. *Gorilla Gut,* she thought as she stared at the bulging protrusion.

Her hands trembled as she pulled her loose maternity shirt down over her head. *Don't do this now,* she told herself. *You have to relax.* She had finished securing her hair back with a Scrunchie when the door flew open and Keith stood in the frame behind the intruding eye of a video camera.

"Hey, Turtle," he playfully taunted, "let's see the turtle shell."

Stacey knew he was trying to dismiss her disquietude with comedy. She wanted to play along—knowing he meant well, but she couldn't produce a smile. He was ignoring the gravity of her situation. Albeit, she surrendered to his request by lifting her shirt high enough to display her jovial belly with a sour scowl poised on her face. Ten seconds of documented history later, he was either satisfied with the shot or recognized his antics were falling far short of helping her relax. She smoothed out her shirt as he shut the camera off and eased his way from the bathroom. She returned her focus to preparing for the challenge at hand. *A few more hours and it will all be over. I'll have my baby.*

Entering the operating room, Keith saw the familiar, large, obstructive sheet draped above Stacey. To the unwitting person, a Caesarean operation would look less like a miracle and more like a maniacal act by the Barber of Seville. After the surgeon opens the stomach, it is necessary for some of the organs to be lifted from the cavity and placed next to the body to allow access to the womb—all being performed while the patient is in a state of complete cognizance.

Keith quickly decided that, this time, he wasn't interested in what was going on beyond the sheet and made a beeline for Stacey's face. Having witnessed some of the surgery during the birth of Samantha, he was no longer intrigued by the procedure.

He started with the customary question, "How are you feeling?" But before long, his tongue became a data processor, spitting out words with the frequency of a news wire at ratings time. He refused to give Stacey a chance to feel the least bit of discomfort as he talked about everything from Christmas to their family in the waiting room, completely skirting the subject of what was happening to her torso.

Stacey heard some of his comments; however she was far from a captive audience as she fought against the bizarre sensations occurring in her stomach region.

Keith pressed on with the tenacity of an auctioneer, fearing his strategy was nothing more than a will-o'-the-wisp. *Distract her whether she wants it or not.*

Soon, fate sought to unveil his premonition as Stacey's complexion paled. "I'm going to get sick".

Keith attempted to alert the operating team, who were busy tending to Stacey, but the adrenalin rushing through his body elevated his voice to a yell. "She's getting sick!" One of the nurses shot him a mitigating glare as she handed him a basin. He held it against her cheek. Suddenly, her head turned awkwardly, her mouth gaped open and yellow-brown disgorgement filled the steel container in Keith's hand. Love and compassion held sway over repulsion as he gently brushed the hair from her forehead and dabbed her mouth with a paper towel. He desperately wanted her to be okay. He wanted this to end. Their countless dates, the years of engagement and marriage all floated to the surface as Keith flooded with emotion. Tears filled his eyes and the words 'help her' formed on his lips, but were never given voice.

"Look what we have here" someone called from under the sheet. "It's a little girl".

"Is she healthy?" Stacey asked the question that was destined to haunt them in the years to come.

An eternity passed as the couple anxiously awaited their response. At last, the words they longed to hear rang in the new parents ears like a symphony. "She looks great."

Keith leaned down and kissed his wife on the forehead, "You have your little girl."

A nurse took the tender infant to a table were heated lamps warmed her tiny body while she underwent a thorough examination during which her length and weight would be recorded. Fear and frustration awoke the sleeping child and with a lip quivering wail Isabella Darlene Cole said 'hello' to the world.

Keith repeatedly paced from his wife to his daughter like a nervous puppy in an attempt to reassure himself that both survived the ordeal unscathed. Soon, a nurse carried the now swaddled Isabella and placed her gently into Stacey's arms. The moment froze in Keith's mind—a snapshot of mother and daughter's first embrace. The moment would be brief as Stacey began to shiver uncontrollably; a condition the attending physician assured Keith was perfectly normal.

The beautiful little girl was passed from mother to father and in the moment Keith held his precious new gift to his chest, his thoughts drifted to Samantha. *Help me watch over this one, will ya, Sam?*

February 6, 2005

The sun offered a blanket of unseasonably warm temperatures as Keith set out to work at one of the nine locations of the restaurant chain he supervised. It was early morning; too early for Stacey and two-year-old Isabella to be awake. Isabella had developed a low-grade fever which persisted throughout the night, but she was now sleeping peacefully.

Keith was sure to have an extremely busy day ahead of him. Super Bowl Sunday always brought out the seekers of snacks and anything deemed a manly, football food. Restaurants, grocery stores and alcoholic beverage establishments would have banner days. The Super Bowl didn't need to be recognized by the politicians in Washington; the game had slowly morphed into a national holiday through ingenious marketing. We, the people, voted—it was time to celebrate.

Keith had volunteered his services to one of the restaurant's managers because he knew the pace would be hectic, having begun his career with the company as a cook himself. Their day of struggling to keep up with the demand of the hungry masses would be a warm-up to the gridiron battle of the game. Somewhere hundreds of crazed fans with war-painted faces in the colors of their favorite team's coat of arms have decided they desperately need secret-recipe, deep-fried chicken for this evening's battle. God bless capitalism over war.

Stacey dressed Isabella in slacks and a black blouse, adding a baby blue headband, complete with bow. Stacey had always prinked for any occasion and was priming her daughter to do the same.

Isabella was given her choice of breakfast which was predictably scrambled eggs and oatmeal—two of her favorites. The toddler ate slowly and finished very little, a clear sign to Stacey that Isabella was still feeling a little under the weather.

While Isabella ate, Stacey called her mother, Darlene Cuozzo, to arrange a meeting at the park. She hoped the unusually warm air might have a positive effect on Isabella's condition. Besides, any child's spirit can usually be elevated by a swing ride in the park with mommy induced propulsion.

Darlene agreed and since the park was close to her house, she walked the few blocks to see her granddaughter play on the toddler swings.

Isabella was thrilled to see Grandma, who did not look or act her age. *Youth is a mind-set and age can be ignored. At least for awhile.* Isabella loved the way Grandma would kick her feet up high and shout as she dangled from the swing next to her. Then Darlene pulled herself up and locked the crossbar behind her knees, swinging upside down like a young gymnast while Isabella shrieked with delight.

Isabella was Darlene's second grandchild. Stacey's only sister, Tanga, and her husband, Don Brown, an entrepreneur and inventor from New Jersey, had a six-year-old daughter named Courtney, who the couple raised in Don's home-state. Courtney had been repeatedly asking Don and Tanga for a baby brother or sister. So when Isabella was born, she was elated. Despite the distance between, the cousins managed to remain inseparably close.

Don and Keith also enjoyed a common bond, distance running. Whenever the Browns visited, the brother in-laws would get together for early morning runs. They also shared in the belief that fatherhood was a huge responsibility and a blessing not to be taken lightly. Shower your children with love and attention and all other activities in your lives will flow through the love of the children.

Stacey was happy to see Isabella remaining active at the park. If the fever still persisted, it wasn't showing any effects on the child. After several hours, they bid goodbye to Grandma and headed home for Isabella's afternoon nap. Keith would be returning home also, and undoubtedly, he would petition Stacey to attend one of the Super Bowl parties being hosted by family and friends.

It was 3:30 in the afternoon when Keith crossed the threshold, exhausted from a strenuous day at the restaurant. Stacey was in the kitchen loading the dishwasher for its next run.

"How's Bella" Keith asked.

"She was still a little feverish this morning but she played hard all day." Stacey replied optimistically. "She just got up from her nap if you want to go see her. She's looking at her books in her bedroom."

Keith took the stairs to the second floor and rounded the corner to see Isabella's usual posture during her book perusing sessions, the back of her head facing him as she leaned over to point and memory read from her favorite stories. Her chestnut brown hair may have been slightly knotted; the standard dress code for a girl of two, but it never appeared that way. It waved through the air as though each strand had been purposefully placed against the backdrop by the skilled brushstroke of one of the great masters. Her chin was a paternal DNA test, cleft like her three siblings before her. Her large brown eyes mirrored her mother's. And the shape of her lips—unmistakably Stacey's.

"Fee-Fie-Foe-Ferd," Keith's twist as applied to the singsong warning of the Jack eating giant, "I smell the blood of a baby bird."

At feeding time as an infant, Isabella would tilt her head back and open her mouth like a baby bird waiting to be fed. And when she began making her first vocal attempts, she even sounded like a bird. Thus, the child had unwittingly provided her father with her own nickname, Bird.

Isabella turned with her beautiful smile that had captured her daddy's heart on countless occasions. "Daddy" was all she could manage in her excitement.

She managed to quell her enthusiasm long enough to catch her breath, but her full lipped smile continued to stretch from pole to pole. "Want Daddy to swing ya" she pleaded with jubilance.

No one could ever confuse pronouns in such an endearing way, Keith thought. *What a blessing this kid is.* "You want me to swing ya" he bellowed back to tease her. "Well alright then, here we go".

It was perfect harmony of father and daughter; a moment of pure enjoyment that seemed to go in slow motion. His arms wrapped around her from behind with his hands joined at her chest, he spun in a little circle as she watched her feet fly in front of her and the world beyond flashing by in a whirl of colors. Isabella laughed until she could hardly breath—'swing ya' with daddy was heaven to her. And to Keith, heaven was a giggling little girl who was nestled safely in his arms.

"I love you, Baby Bird."

Keith turned onto Wilson Street while Isabella entertained herself in the back of his Subaru wagon. She continuously told herself that she was going to Nanny's, as though repeating the mantra would somehow help to precipitate the event. Suddenly the child's attention was averted as she shouted with delight, "Moonie."

"What are you talking, Bird. The Moonie comes out at night."

"Moonie," she insisted.

Keith pulled the car tight to the curb and surveyed the clear, winter heavens. The sun hovered above the western skyline, kissing the trees in its descent. A handful of puffy white clouds drifted slowly against the blue sky. "Honey, its daylight and . . ." He cut his comment short when he found the moon's ghostly placement high above the southern horizon beside the billowing clouds. His daughter's love of the moon had drawn her eyes to see what he had nearly missed.

"How did you see that up there, Sweetie?" he asked.

She responded with a single word, "Moonie!"

It was the wondering eyes of a child seeing the world in a way in which he had forgotten. Keith had committed the greatest sin against the pursuit of visual beauty and awe—he had grown up.

Wilson Street had undergone little change since Keith's childhood. On the south side of the street, the Wilk's house had received a vinyl-sided facelift from the new owners—and a new porch to replace the original wrap-around style. The Brown's had fortified their yard with a chain link fence. On the north side, the old Hartley place now had an enclosed porch, but apart from that; thirty years had made little difference. The white, aluminum-sided, double-house where Joyce Cole raised Keith and his seven siblings remained unchanged, minus a few pieces of slate that occasionally dislodged from the roof. It was a good place to grow-up and a good place to return. It was here that Joyce, wife and mother, had transformed into the family icon known as Nanny. An epitome of womanhood, Nanny was a caretaker of immeasurable strength and perseverance. With every child she bore, she wore a dozen hats; maid, nurturer, chef, short-order cook, nurse, referee and teacher among others. But that responsibility came with a purpose that steeled her against the turmoil that often accompanied her life. From children came grandchildren and from grandchildren, great-grandchildren. Isabella was Nanny's twenty-sixth and final grandchild to date, a number that would engender an astonished smile even from Nanny.

It had been almost two years since she had reluctantly agreed to babysit the infant while Keith and Stacey worked. Nanny was more than up to the task; however, the sixty-eight-year-old widow had reared her children and now enjoyed an independence that she was hesitant to relinquish. But the couple had no one else they could rely upon—which tweaked her celebrated sense of compassion. She complied with their request and later came to realize her decision enriched her in ways she never imagined. Isabella filled an emptiness left by the passing of her husband, Ralph, who suffered a massive coronary and died before the paramedics could respond. In the time before Isabella, Joyce would often set out in her car to merely escape the pain of Ralph's absence. The kitchen table where they shared coffee and the bedroom where they nestled in one another's arms were haunting three-dimensional snapshots of happier days. After thirty-nine years of marriage she was worse than alone—she was the only remaining half of a single entity. But now that entity was made whole again through a bond with an imaginative little girl who insisted that Joyce was more than a nanny; and Nanny was more than a caretaker. Nanny was a friend; a dance partner and a marching companion of Captain Feather Sword, plucked straight from the toddler's beloved Wiggles television program. Through the passing of time and the singing of songs, Nanny soon referred to Isabella as her ninth child, and, from Isabella's account, her best friend.

Nanny and Isabella's bond brought about many trying times for Keith and Stacey. Isabella refused to go home. A bitter-sweet battle waged at nearly every appointed pick-up. Isabella's battle-cry of "I want to stay with Nanny" proved a source of frustration and rejection for the over-worked parents. They had found the best possible childcare and, as a result, isolated themselves from their daughter.

This visit promised to be no different, Keith was a tolerated interloper waiting to snatch the good Captain Feather Sword from her first mate as he sat on his mother's couch while the two played upstairs. He could hear Isabella dictating every move to Nanny, and Nanny laughing at her willfulness.

After an hour, Nanny came downstairs and told Keith that Isabella was jumping in her crib when she suddenly stopped and said, "Nanny, tummy hurts." Keith decided it was best to take her home. Isabella offered no resistance as Keith dressed her for the cold. Nanny kissed the child goodbye and watched out the kitchen window as Keith carried Isabella to the car.

February 7, 2005

Stacey awoke Keith from a restless sleep at 1:30 am. Isabella's fever had worsened and Stacey was exhausted from trying to comfort her. He went into Isabella's room while Stacey returned to bed for some much needed sleep.

Keith cradled Isabella in his arms as he sat in the nursery's rocking chair. He could feel the heat emanating from her little body. The doses of Tylenol they had administered to her earlier were having little, if no, effect.

Isabella held her favorite *Blankie* to her nose and breathed into it. The receiving blanket was one of a set they had purchased at a discount department store; predominantly blue and emblazoned with teddy bears. She was a master at self-soothing, but she needed certain tools to perform the task. *Blankie* and Binkie, a colorful pacifier, were her two essentials for going to sleep or weathering any illness. When Blankie got soiled, Keith and Stacey had to rotate the two blankets in order to launder the original.

As the minutes passed, Keith's attempts to comfort her into a peaceful sleep fell hopelessly short of their goal. Previous experience with false alarms made him reluctant to rush her to the emergency room.

To a couple with a young child, a fever or a tummy ache can lead to panic which can oftentimes lead to embarrassment. Isabella already had a history of stomach problems. One such flare-up sent the couple into the ER with Isabella screaming at full lung capacity. No sooner had they arrived, Isabella quieted down. The examining physician gave her a simple dose of apple juice. And after Isabella enjoyed a much relieving poop, Keith and Stacey departed for home, humiliated and vowing to be more careful about hospital visits in the future.

The night drudged on and Keith became more concerned with each passing minute. At 4:30 am, he was able to place her in her crib. He grabbed a pillow and lay on the floor. He listened as Isabella kept moaning. At first, he hoped she would stop and fall asleep but it was soon apparent his hope was not helping the child to rest.

He spent the next hour in a sleep deprived cycle of holding her hand and then attempting to lie on the floor. But the moaning continued until his fear overrode his caution.

"Stacey." Keith gave his wife a slight shake to awaken her. "She hasn't slept at all and she keeps moaning. I'm really worried about her. We need to do something."

Stacey wearily rubbed the sleep from her eyes. "Let me check her temperature." It was 104 degrees. "Let's take her to the emergency room."

They sat in the waiting area of the Williamsport Hospital's ER for about an hour after a nurse had checked Isabella's vital signs and administered some Tylenol. On the drive to the hospital, Keith had called Nanny to notify her of Isabella's condition. The wait was an eternity of rising anxiety. Keith shot an occasional glance at the faces of strangers who had the misfortune of sharing the wait, as he wondered what perils may lay ahead for some unsuspecting family. He had been on the receiving end of more than his share of crisis situations. Samantha's accident, his father's heart attack and more recently his eldest brother's struggle for life after suffering third degree burns over 70% of his body following an attempt on his life. For Keith, television shows that portrayed reality in the ER were no longer viewed with a lounging sense of obtuseness. Not only could these things happen to him, they did. The memories and emotions of reality were far too fresh to endure the suffering of those on the screen.

A nurse entered the hallway adjacent to the waiting area and called Isabella's name. She led them into the treatment area which consisted of a work station pressed against the western wall isolated by a large surrounding counter. The rooms were separated with thick, pinstriped curtains that ran along a tracking system in the ceiling. Each dock had a sink and a utility desk with most of the instruments necessary for the treatment of traumatic injuries. The nurse gave Stacey a gown for Isabella and reassured them the doctor would be with them as soon as he could.

As they waited for their turn with the attending physician, Isabella was quiet, clinging to her mother as a wet autumn leaf clings to a tree. With a pacifier in mouth and her favorite blanket in her hands, she rested her head on Stacey's shoulder. At last, the curtain drew back and a tall, middle-aged man stepped into the room.

"So, Isabella is complaining of stomach pains and has been running a fever?"

"Yes." Stacey, being a registered nurse, was the obvious candidate to handle the situation.

"I want to check her stomach." The physician lifted Isabella's gown and began pressing on her stomach. Isabella moaned. "How long has she been experiencing the symptoms?"

"Since yesterday, but her fever has gotten worse since last night."

He listened to her belly with a stethoscope and then listened to her chest. "I'll come around to the other side. I have to check her back."

Isabella was groggy and almost unresponsive as she sat on Stacey's lap. The whites of her eyes clouded with the grey of an approaching storm.

"How did this happen?" The physician asked as he pointed to a spot on Isabella's back.

Keith drew closer to see the quarter sized, reddish purple bruise on Isabella's lower back. At first, Keith and Stacey responded only with a shrug and a quizzical glance at each other. *Did this happen on your watch?*

"I didn't know it was there." Keith said.

Stacey concurred.

"Okay. I'm going to run some blood tests and have her catheterized for a urine sample. Her fever isn't responding to the Tylenol, so that brings enough of a concern to have her transferred to Geisinger Medical Center in Danville as a precaution."

Keith's heart thumped against his ribs, threatening to break itself in the process. Geisinger Medical Center was the premier health care facility in Northeastern Pennsylvania, but it was also the very hospital where Keith watched helplessly as Samantha passed away. *Not there again.*

The ER physician sensed Keith's concern. "She is in no imminent danger." he said, reassuringly. "A nurse will be in shortly to run the catheter."

He disappeared through the curtain like a magician escaping a cynical audience. Keith turned his attention to Stacey, "When do you think she got that bruise?"

"I don't know."

The idea of the doctor calling childcare services flashed across Keith's mind. The bruise looked like the result of a strike and the fact the doctor felt strong enough to question it made him apprehensive.

When the nurse came in to catheterize Isabella, Keith stepped outside with his cell phone to call work. He had already decided to take the day off. With that out of the way, he turned his attention to calling Nanny. As he dialed the number, he released a deep sigh. He had to inform her of Isabella's condition, and he knew she wasn't going to take it well.

He finished his calls and flipped his phone shut. The haunting similarities between his daughters created a cold chill in the pit of his stomach. Samantha was 27 months and 3 days when she passed, and now, Isabella, at 26 months and 13 days was being transferred to the same medical center with questions of her condition remaining unanswered. He suddenly recalled a conversation he had with his eldest son, Jesse.

They were driving to the local shopping mall on a pre-Christmas jaunt, when Keith confided in him.

"I feel like we just have to get through February and Isabella will be alright," he had said.

Isabella was not sick at the time of the conversation, but Keith had been carrying a feeling of dread since the start of winter. Keith tucked the memory away and returned to the emergency room as Stacey was carrying Isabella toward the restroom.

"They're having trouble running the catheter," she informed him. "I want to try to get her to go on the potty so they can get their sample that way." Stacey spoke as she fought back the tears that welled in her eyes.

Inside the restroom, a large mirror was hung on the wall beside the toilet. Stacey lifted Isabella's gown to place her on the seat, but froze as she caught a glimpse of her child's back as it reflected in the mirror. Dread coursed through her body and spilled a thin stream of tears down her right cheek. She shot a glance at Keith. His widened eyes gave her the confirmation she needed. He was seeing it too. The quarter sized bruise now covered nearly half Isabella's back and had darkened to a deep purple. Stacey scooped Isabella into her arms and rushed back to the appointed ER dock with Keith following on her heels.

"Get the doctor," Stacey instructed him as she took Isabella back behind the curtain.

Keith ran to the work station, his mind a raging flood, taking all optimism in its wake. His voice released in a shrill holler, "Doctor, you need to come and look at her back, the bruise is bigger!"

Keith led the ER physician through the curtain into Isabella's dock. Beyond the divider, Stacey sat in a stool as she pressed

Isabella tight to her bosom. Isabella pressed *Blankie* tight to her nose.

After re-examining Isabella's back, the physician repeated the mantra, "She is in no immediate danger." Then he added, "The helicopter is on the way."

Stacey caught sight of another patch of discoloration near Isabella's right groin. Stacey's nursing experience was beginning to click in gear. She was now thinking the worst.

Quiet surveys of the room gave way to tight-faced glares directed toward the nurse's station and then to caged-tiger pacing as no sign of the promised helicopter came forthright. Minutes were becoming an intolerable collection of lost opportunities to save their daughter from whatever it was that was now attacking her tiny body. Keith looked back at Isabella. Her complexion was dull and pale behind the colorful plastic of her pacifier they called her Binkie. A small stream of red liquid began to trickle from beneath Binkie's seal.

"Is that blood" he asked Stacey.

Stacey looked and said "No. It's not blood."

A man and a woman entered through the curtain, each dressed in a blue jumpsuit, the emblem of a helicopter was sewn to the breast pocket and beneath, the words "Life Flight" was embroidered in red. Before they could introduce themselves Isabella gagged. Her mouth flung open and a mixture of blood and apple juice sprayed the floor in a violent retch. The thick, crimson arc of vital fluid was cast out with such force that only a few spatters hit her gown before it splashed upon the floor. The air was sucked from the room leaving a cold, arid chill that crawled between the hairs of their flesh.

"You waited too long!" Stacey screamed as her entire body quivered under the stress. Overwrought, dread had now robbed Stacey of her tears leaving her little physical reaction beyond uncontrollable trembling and a vacant, silver-dollar gaze.

Keith watched as the very color of life drained from his daughter's face. He had witnessed this before. It had been Samantha. And now it was Isabella, dying right before his eyes. The natural glow of animate existence, ever present on the skin, goes unnoted until the pallor nothing of lifelessness descends to waxen the complexion of the human face. The sight of death is unsettling; the sight of dying is horrifically indefinable. Samantha had been in a coma, but Isabella was awake and staring directly at him. The precious full-lipped little girl who, less than twenty-four-hours-ago, had squealed the request, "Want Daddy to swing, ya." was far more ill than he had expected.

Panic clung like mud to Keith's feet, sucking him to a fixed position; his eyes locked helplessly on Isabella's colorless little face. The attending staff seemed to be gripped with the same paralyzed fear that held dominion over Keith's body. No one moved to help the child.

For a moment, only Stacey provided assistance to Isabella as she tried to wipe the liquid off her toddler's face. Isabella's stomach lurched, her mouth gaped and more blood and juice showered the floor.

Isabella's eyes met Keith's in a frantic stare. She didn't recognize the expression on her daddy's face, but what she sensed in him confused her. This was Daddy the protector; Daddy the hero; Daddy who 'swung ya'; Daddy who made her laugh and kissed her boo-boos; Daddy who made things better. But now,

Daddy looked lost and frightened. She desperately needed him to rescue her, so in the tender voice that had so often confused pronouns, Isabella rasped. "Daddy, fix it."

At last stumbling forward, Keith rushed toward his daughter, fighting against crumbling to the ground, as he released an anguished cry, abashed in sheer helplessness. "Oh my God!"

Danville, Pennsylvania

Interstate 80 exemplified Pennsylvania with its expansive landscape of rolling hills and dense growths of pines, maples and hemlocks on both sides of the eastbound and westbound lanes. The air was crisp and the sky was overcast, but all went unnoticed as the Subaru sped east, destined for the Geisinger Medical Center in Danville—an excruciating forty minute trip from Williamsport for the parents of a child clinging to life. Keith, still in shock from what had transpired only moments ago at the emergency room, drove with his eyes fixed despondently on the road ahead as he made several cell phone calls to alert family members about Isabella's condition.

"My beautiful little bird." Stacey repeated the words from the passenger seat as she wept uncontrollably. Her face, etched with her suffering, was pallid and streaked, and her swollen eyes looked as though she lost another day of sleep with every passing minute.

They could barely speak to each other under the intense pain. Keith knew he should comfort Stacey, but he could barely hold himself together. His child was dying and his wife was crumbling, and there were no words or deeds left in his fracturing heart to help either one of them.

The irony of traveling the same route from Williamsport to Danville for a second daughter fighting for her life was all consuming. In his heart, he feared the worst. In his mind, he couldn't comprehend being put on Earth to suffer the deaths of two daughters so close in age, separated only by the fifteen years between each tragedy. Yet his biggest fear was close to becoming a reality.

After making the necessary calls, he drove with his eyes on the road and his mind on Isabella, and the gentle way she had made him a better person. Every moment of fun and every smile she wore began running through his mind in no particular order. Like a slide show out of sequence, he relived as much of Isabella's life as his tortured heart would allow.

He recalled the first time he gave her lima beans—she spat them out, telling him for weeks, "Isabella no like a da lima beans." He thought about the way she walked for the first time; arms bent at the elbows, fists clenched and moving forward and backward with every step trying to keep her balance. It was like some beautiful dance she had invented, just for her parent's viewing pleasure.

He thought about her love of music. When she was having trouble sleeping as an infant, Keith would play a song on his stereo and walk around the family room with Isabella cradled in his arms as he sung the lyrics from a *James Taylor* song or oftentimes *These Eyes* by the *Guess Who*. She would stare, wide-eyed, at her Daddy until eventually succumbing to sleep. It was a lullaby technique that made Keith proud.

Another song she had grown quite fond of was *American Woman*. Not because of the love repudiating lyrics or the punchy guitar riff that inspired countless air guitar solos, it was the simple inarticulate "huh" during the intro. She would emulate this for family at Keith's request as she rocked her head like a miniature rock star.

As of late, the Wiggles ruled the airwaves at the Cole house and Keith had come to know every Wiggles song by heart. Isabella would yell each time a favorite song ended, "Wanna see again." A phrase she had grown accustomed to saying when she wanted to hear or do something again. "Wanna see again." Now Keith wanted to see it all again. He wanted to gather all these thoughts together and pull them close for safe keeping. It felt like his chest would soon explode and the memories of his child would flood the car and drown him.

But all of the beautiful thoughts came crashing down as he suddenly pictured her face just after she vomited in the ER. That sudden loss of pigment that meant the life-force was leaving a living, breathing human being; a sight too harrowing for human eyes and too agonizing for the human heart.

The smiles and the singing were gone, now replaced by an awful reality.

The look on her face as they placed her into the car seat to prep her for the helicopter ride had been pitiful. She had shook uncontrollably and attempted to cry, but was so weak it sounded more like a wounded baby animal than a little girl. Keith remembered thinking, she didn't know why this was happening to her or why her mommy and daddy didn't just take her home and hold her in her rocking chair.

The good and the bad enveloped Keith and Stacey as they drove not knowing what awaited them in Danville. By now,

Isabella had already arrived frightened, alone and enervated, watching strange faces pass before her.

As a registered nurse, Stacey witnessed the same frightened expression from patients she had treated. But this time the child was her baby girl. Stacey's body quivered at the thought. No matter how hard she had to fight, Stacy was determined that Isabella would receive the best possible care. Anything less would be unacceptable. Stacey was tattered and near a faint, however the fear of losing her daughter was taking root in her instincts and steeling her resolve—the loss of precious time at the ER would not be repeated.

An expansive highway peppered with gas stations, casual dining establishments and fast food restaurants disappears at the edge of the quaint town of Danville. A left at the traffic light allows you a quick glance at the center of town before the scenery transforms into Anytown, USA as bi-level homes with widow-peak roofs hide behind shady maples along either side of Route 11. Soon, an oddity appears on a hill to the north, the Geisinger Medical Center, looking like a metropolitan facility that had been dropped into the middle of a rural neighborhood as it looms over the treetops to survey the small town it occupies.

Danville, the seat of Montour County, the smallest county in Pennsylvania, was founded in the late eighteenth century by William Montgomery who named his settlement after his son Daniel. After becoming home to the first American T-rail mill, the town became a pathfinder during the iron era of the nineteenth century. The iron railroad spread across the country, the town prospered and individual entrepreneurs amassed vast caches of wealth as a result of their iron foundries. One such iron magnate was George F. Geisinger, whose widow, Abigail, at the staunch age of eighty-five, founded the George F. Geisinger Memorial Hospital. Abigail recruited Dr. Harold Foss as her surgeon-in-chief, and together, they pioneered the concept of group practice into what is now know as the Geisinger Medical Center. At the time of Abigail's death in 1921, her facility serviced 156 communities and was known as the largest healthcare center in the United States. By the new millennium, the hospital of seventy beds had swelled into a healthcare center that boasted over four hundred hospital beds.

Keith turned left onto Academy Avenue. The road rose as the car approached the eastern corner of the eight storied, sandstone building which stretched out to the left. To the right, a parking lot, covering more than twice the acreage of the main building, sprawled westward before wrapping around the north side of the facility where the entrance to the emergency room awaited them. Immediately upon entering the building, Stacey and Keith followed a nurse down a dimly lit hallway to a waiting room. As he approached, the sight of his two sons chatting outside the door granted Keith a temporary moment of solace. He needed them close, now more than ever. Fifteen years ago, they had been present during the drowning of their sister, Samantha, and every passing day since, they witnessed the burning pain in Keith's eyes. Jesse and Shilo seldom discussed the loss of Samantha with Keith, an avoidance he attributed to their desire to help him move on, but he knew the loss of their sister tortured their hearts. And now an agonizing turn of events brought the three men together for

Isabella. Keith held them tightly, attempting to draw upon their strength.

Stacey stepped through the door into a small waiting area where family and friends had already gathered, anticipating word of Isabella's condition. Stacey was dazed as she described some of the events of the past few hours. The members of her and Keith's family listened intently, trying to gather any hint as to the cause of Isabella's bizarre symptoms. Keith entered the room and flopped down into a chair, his eyes red and puffy and his hands maneuvered with a slight tremble. Five minutes passed with no news of Isabella. Keith and Stacey made the decision to approach the nurse's station.

A pleasant brunette greeted them and agreed to take them back to visit. With questions surrounding the possibility that Isabella was afflicted with a contagious illness, they were required to put on hospital gowns and masks.

The intermittent clicks and beeps of the blood pressure apparatus and heart monitoring equipment added a disconcerting presage to the thick, almost palpable atmosphere of dread within the room. An IV drip had been started and an oxygen mask covered Isabella's face. With her breathing dangerously erratic, the oxygen was set on one hundred percent. *Blankie* remained faithfully at her side.

Isabella's eyes were half closed as Keith and Stacey held her little hands and began to let her know that they were with her.

Stacey looked at the chalky face of her only child through rheumy eyes, "Mommy and Daddy are here, baby bird. Rest your little head now. Everything is alright."

Isabella talked as if she were in a dream state, mumbling incoherently. "Elmo has his tap shoes on." Occasionally, she would smile as if she were watching an amusing show on the television in her bedroom.

"She's heavily sedated," the nurse explained.

Keith lowered his face and kissed her brow. "At least she's not scared and suffering."

"Do they know what's wrong with her yet?" Stacey asked.

"No," the nurse replied. "They aren't certain, but I know they were talking about the possibility of bacterial meningitis."

As soon as the nurse left the room, Keith began to quiz Stacey on the disease. "Is bacterial meningitis bad?"

"Yes." Stacey snapped.

Keith suddenly felt silly for even asking the question. He had no idea what bacterial meningitis was, but he had heard the term enough to know it wasn't good.

"What's the survival rate?" he asked next.

"I don't know, but it isn't very high," Stacey said. As a nurse, she was always more than happy to answer medical questions for family members, but now her daughter was also the patient. She didn't want terminology, numbers and statics running through her mind. She was a mother now. She had the right to worry as a mother. Today she was not the nurse. Keith wanted reassurance from her, but he didn't realize she needed it too.

Keith had a million more questions, but kept quiet, sensing his wife's irritation.

Twenty minutes passed. The couple focused their attention on reassuring their baby, but longed for someone to reassure them. A young ER doctor entered the room and addressed them with eyes that held more compassion than answers.

"We still don't know what is wrong with your daughter. We're running blood work to try to determine the cause. This much we know; you have a very sick little girl here. Her body is in shock from this infection. We are giving her antibiotics and we have her on oxygen to help her breathing. We will do everything we possibly can to get to the cause."

"The nurse said this could be bacterial meningitis. Is that true?" Stacey asked.

"It is a possibility, but there are some things here that are not consistent with that diagnosis. We're just not sure, yet. We will give you a few more minutes. Then we have to ask you to leave. The nurse will take you to the waiting area for intensive care."

They kissed their little one and bid tearful goodbyes before being led back to the waiting area to give their families an update. By now, Keith and Stacey were growing tired of telling the story of Isabella's illness and how it all transpired. As new family members arrived, they struggled through the pain and frustration to politely update each.

Following several minutes, a nurse came into the room and requested everyone to transfer to the pediatric intensive care waiting area. The room was much larger with windows running from floor to ceiling that overlooked the north side of the parking area. The cold, indignant sky offered no sunlight, and no optimism. The clouds, smeared in watercolor grey, hung low in the sky as they rushed northward, appearing as though, if raised on their toes, a person of average height may be able to stretch far enough to touch them. The sidewalks leading away from the hospital offered no easy path to home; at least not with a healthy, more vibrant Isabella.

When Keith's eldest brother Ralph arrived, he asked Keith to take a walk with him. Keith reluctantly agreed, knowing what Ralph wanted, but not wanting to talk. He wanted Keith to say, "I'm going to be alright. I can tough this out." Since the passing of their father, Ralph proudly bore the responsibility as the head of the family, and as a result, felt an innate responsibility for Keith's welfare.

Five minutes had passed since Keith and Ralph departed for their somber stroll around the expansive grounds of the facility when a stout female doctor with short curly hair came in with another physician in tow. She greeted Stacey, recognizing her from previous encounters within the past hour. The in-tow doctor was a new face—a man, who appeared in his late fifties that she identified as Dr. Romanowski. The man's expression was solemn. The room held its collective breath. But what he had to say would soon knock the wind from everyone in attendance.

It seemed colder than the thermometer was reading as Keith walked along the sidewalk that surrounded the Geisenger Medical Center. Even the slightest breeze snapped at his skin and quickly numbed his ears.

Ralph was silent for awhile and Keith allowed him the time to gather his thoughts.

Keith knew what Ralph wanted to hear—needed to hear. Ralph needed to hear that Keith was okay; that he was stronger now than he was fifteen years ago; that he wouldn't crumble, even if Isabella did not survive. Ralph's concern went beyond the love of a brother, he possessed a soul with an intrinsic need to rescue others, and he was well equipped for the job. He was a hero who possessed the tools to accomplish that which others lacked the fortitude to endure. Sculpted in granite nearly from birth by a father who goaded him relentlessly, Ralph became a human bull not given to the allure of a red cape. Any man unlucky enough to stir his ire became a matador in the direct course of an intrepid charge, and no flashy suit or diversionary cape waving could stave off their mistake. Ralph was the toughest man Keith had ever known, and

one of the most resourceful. But within the stone exterior was a
hearth where family could always be warmed.

Not long after Samantha's death, Keith moved out of his
parent's house into a small place of his own. To say the bi-level
home was modest was a gross understatement, but it was his
with all the space its four rooms could provide. The house was in
need of repair, and Ralph, who was good with a hammer, would
graciously stop by on occasion to do minor fixes. On one such
stint, Ralph noticed Keith's home was not the only thing in need
of repairs. Grief was taking a physical toll on Keith. His weight
loss, his ashen face, the distant, vacant look in his sunken eyes was
all too apparent. Ralph persuaded Keith to attend a group therapy
session being held in Danville. The support group was intended for
parents who lost babies to miscarriages, though Keith was warmly
greeted. When it was Keith's turn, he told the story of Samantha
with a watery voice. He imparted the sublimity of parenting such a
gentle spirit, the joy of her life and the despair of her death, leaving
no heart in the room unbroken and no tear unwept.

Following the session, they drove back to Williamsport with
a stronger bond than they had ever known. Keith reassured Ralph
that no further therapy was needed. The release of emotions had
helped, but the discomfort of opening his heart to strangers made
him vow to keep his grief private.

Now, as they walked the grounds of Geisinger Medical Center,
Ralph remained private regarding the true impetus of the walk.
Knowing that Keith owned a handgun and fearing he would
attempt suicide, Ralph had already asked Jesse, Keith's eldest
son, to secretly remove it from the home and place it in his care
until the situation passed. Albeit, he needed to test his brother's

resilience to the overwhelming anguish which, surely, swelled within him.

Later, upon finding out about the weapon, Keith would tell his son, "Jesse, I could never do that to you and Shilo. Not ever!" A fact he knew from experience. Words that may have afforded Ralph some comfort if spoken in advance.

Today, words to ease Ralph's apprehension were not coming easy.

"This is so strange," Keith said. "I can't believe this is happening again."

"The family is worried about you," Ralph said. "You got to hold it together."

"I feel like I did something terribly wrong; like I'm being punished."

"It's nothing you did wrong, brother," Ralph said, trying to reassure him—trying to rescue him. But Keith already knew the answer. This wasn't happening to him. It was happening to the sweet little girl fighting for her life in the intensive care unit.

As Ralph continued to talk, Keith recalled a similar walk that Ralph had requested less than an hour after Samantha died. It was the same kind of zombie saunter they were embarking on currently. They had spoken without saying anything; moved, yet stood still in time.

Fifteen minutes passed as Ralph spoke and Keith's mind wandered to the precious palaces of reflections his daughters had created for him. Rising and falling, his emotions Ferris-wheeled through various mind-theatre clips of birthdays and holidays, first steps and missteps, until finally Keith spoke up, "I want to go back inside, Ralph. I need to hear some news. And I don't want to be missing for too long in case I'm needed."

As they exited the elevator and rounded the corner to the waiting area, Keith saw the expression on the faces of relatives and friends—their lashes moisture-swollen, their ears flushed from elevated blood pressure and a woeful apprehension twisting their darkening faces. Some averted their focus to avoid making eye contact with him. Others seem to be awaiting his reaction to whatever loathsome news that awaited him. He suddenly felt embarrassed instead of fearful, and for that he hated those looks. He hated the expectations as though he should prepare some thoughtful catastrophe acceptance speech, or become the captain who steels his jaw to the oncoming storm as to receive the final ten-story tidal wave that will overturn his ship. Keith braced himself. The wave was coming.

Stacey met Keith's gaze with a look of anguish. "The surgeon was just here. He said Isabella has gas gangrene."

Dr. Romanowski, an experienced surgeon, had been around the operating room long enough to recognize Isabella's infection. The blistering of Isabella's back and groin areas, the discoloration and thinning of her flesh were tell tale signs he had seen before. He knew there was no time to waste. Isabella was infected with Clostridium septicum, a bacterium in the same class as botulism. In Isabella's case, which was a skin and soft tissue infection known as clostridial myonecrosis causing gas gangrene, the disease was progressing rapidly. Typically acquired through septic absorption—the bacteria enters the body through an open wound and once beneath the skin, the anaerobic germ goes to work in its oxygen absent environment, multiplying and releasing alpha toxins that result in localized death of muscle cell fibers. The diseased muscles soon appear gangrenous and black. The cells lining the blood vessels begin to die causing vascular permeability, allowing the blood to seep through the vessels. The blood appears beneath the skin along with the black of the underlying dead tissue. The blood can also be absorbed into the stomach and vomited. More tissue dies as the infection spreads. The diseased tissue is deprived of oxygen which causes vascular insufficiency. There is a loss of blood supply to the heart and the host's blood pressure begins to drop causing an abnormally slow heart beat. The body then goes into shock. The toxins become systemic, spreading through the entire body and causing organ failure which ultimately leads to death. In cases such as Isabella's, all of this occurs within a few days, or even a few hours. In these fulminate cases; the only treatment is immediate surgical intervention, as antibiotics have little effect without it.

Gas gangrene could easily kill someone in a very short period of time if the infected areas were not surgically removed followed by heavy doses of antibiotics.

The disease was aggressive and needed to be countered by equally aggressive treatment. Isabella would undergo a surgical procedure known as debridement, the removing of the infected muscle tissue and skin. This infection loved non-oxygenated areas of the body, so it would be a battle to try to keep it under control. If any of the bacteria were left, it would continue to spread. It was not the time to worry about the cutting that had to be done to save this child's life.

"He needs to operate now to remove the infected muscle and tissue." Stacey explained to Keith. "They need your signature for the surgery. He said she may not live and if she does, they may have to amputate her right leg." Stacey walked away, leaving him frozen in disbelief.

Keith made his way back to the intensive care desk, walking unsteadily on wobbly legs. He signed the necessary paperwork and stumbled into a smaller waiting room to sit alone. Stacey entered a short time later; behind her was her mother, Keith's son Jesse and his girlfriend, Tia. They all stood in the hall as Stacey and Keith sat in the room staring at the bland walls.

"Stacey, what the hell are we gonna do?" Keith finally managed.

She never answered.

Keith and Stacey were allowed to see Isabella briefly as they wheeled her bed to the operating room. Accompanying them was Isabella's Nanny, who was able to hold Isabella's hand for a brief moment.

Nanny spoke through quivering lips, "Nanny's here baby girl. I love you."

Isabella mumbled "Nanny." She opened and closed her eyes several times.

An elderly nurse told them that Isabella had asked her to sing 'Rock a Bye Baby.'

"It broke my heart." the nurse explained through sodden eyes. "I'm a terrible singer, but I sung it for her anyway." Keith surmised that Isabella thought the nurse was her beloved Nanny, who had sung the lullaby on hundreds of occasions to coerce the child to sleep at nap-time. 'Rock a Bye Baby' and 'You Are My Sunshine' were Isabella's favorite Nanny songs. Her voice, while not always in tune, became an angelic, legato refrain shrouded in the nurturing love that emanated within its breath.

For Isabella, the song was the comfort she needed—the reassurance that her life would still be the same when she awoke. It was the same exigency that causes adults to rock themselves or to hum when stress becomes overwhelming. Keith could recall the reassuring sound of his mother's hum from his childhood—and how much he longed to hear it now. It was the universal, inarticulate message that promised everything will be okay.

Keith's mind raced back in time to January of 1990, when he stood in the pediatric intensive care unit of the very same hospital with Lori. Their twenty-six-month-old daughter, Samantha, was suffering from brain swelling after her accidental drowning. Keith and Lori had kept vigil all night, and it was approaching dawn. The doctors had done everything they could to save her. As Keith watched them working on her little body, he saw the burns from the heating blankets in which the paramedics had wrapped her. Desperate to ease her suffering, he asked the staff, "Can't you just leave her alone?"

The doctor had replied "It's alright now. She's dying."

"I can't be here," Keith said as he rushed from the room. A nurse followed at his heels.

"If you don't go back in there, you'll regret it for the rest of your life," she had said.

Keith turned around without a word and went back into the room to see Lori holding Samantha, crying "My baby, my baby" as Samantha's skin turned pale and her lips turned blue.

And now, he couldn't help but feel he had reached the same moment with Isabella. As quickly as she had appeared, they wheeled Isabella away, leaving him to wonder if he would ever see her again. *How could lightning strike me twice like this? Were there warning signs like before? Signs I may have missed?*

As a young boy, Keith had endured two dreams of being struck by lightning, dreams which left him with a fear of the nature phenomenon. He was sure he would be struck one day. Then as a young man, he had another dream. He was looking out his parents living room window when he witnessed a little girl get struck by lightning. The electrical blast knocked the girl from her shoes and frightened Keith from his sleep.

After Samantha died, he couldn't help but wonder if there was some significance to the dream. Now, it seemed even more eerie to him.

If dreams were in fact presages to future events, one which was experienced by a co-worker was equally disturbing. The co-worker was a young woman who had always told Keith of the propensity for her dreams to become reality. Following his divorce, she described a dream to him in which Samantha and his sons were lost in the woods. She ended her story with a portentous, "I thought you should know."

Keith related this to Lori half-heartedly. She told him, "Don't worry. I won't let anything happen to the kids."

Even beyond dreams, the signs of Samantha's demise seemed to be prevalent after the fact. On a trip to the New Jersey beach, Keith had scrawled the names of his children in the sand. And while Lori caught the markings on film with the couple's camcorder, a wave swept in and washed away Samantha's name, leaving behind the names of Jesse and Shilo. While Keith poured through old video tape of his departed daughter, he came across the scene and was startled by what had been captured. Now, he was sure God had tried to warn him about his children. But he just didn't know how to listen. Only after Samantha died, had He managed to gather Keith's attention.

This time had to be different. He decided at that moment, no matter what happened from here, he couldn't accept anything less than Isabella coming home again.

"God, let Isabella live. Take me," he prayed over and over.

"Use her to show the power of prayer. Heal her Father. It's already been one daughter too many. I can't do it again. Please, don't make me do it again."

The stout woman sitting across the table from Keith and Stacey tried to smile as she drew in a deep breath. "You have a very sick little girl," the couple heard for the second time. The walls of the small conference room seemed to close in on the pair as they tried to read the expression on Dr. Wilson's face for any signs of hope. Dr. Wilson, the head of the pediatric intensive care unit for the Geisinger Medical Center, had the unenviable task of speaking to Isabella's parents following the debridement surgery.

"She came through the surgery alright," Dr. Wilson continued, "but she is a long way from being out of danger. Isabella's surgeon, Dr. Romanowski, will be in momentarily to discuss where we go from here. But I wanted to take the chance to tell you, he feels strongly that your daughter needs to be transferred to Pittsburgh's Children's Hospital. Isabella has a clostridium infection, and this particular bacteria flourishes in non-oxygenated areas of the body. He feels the best course of treatment is to bombard the body with oxygen and antibiotics. Pittsburgh has a hyperbaric chamber that can perform this treatment. The pressure of the chamber will force the oxygen into her system and theoretically stop the bacteria from spreading. But I must caution you, there is no hard evidence to substantiate the treatment's effectiveness. Quite frankly, Dr. Romanowski is the only physician involved in your daughter's case who believes this is a necessity. We simple don't have enough evidence that tells us hyperbaric treatment will work for her condition. That means the decision will be yours."

"What's our alternative?" Stacey asked.

"We'll treat her here." Dr. Wilson said, confidently.

The door to the conference room opened and Dr. Romanowski entered. The tall, lean man took a seat beside Dr. Wilson.

"I know you're anxious to hear about your daughter's condition, so let me get right to the point." Dr. Romanowski said. "The good news is she's still alive. We have to be thankful for that. I had to remove a large amount of muscle and tissue from the lower right half of your daughter's back. There was also a large section of dead tissue and muscle that had to be removed from her right leg; around the groin area . . ." He traced an area of his own thigh to show approximately where the child had lost muscle tissue.

"I have to tell you—your daughter may not survive. She is definitely not out of the woods," he continued. "I believe the best place for her is Pittsburgh's Children's Hospital. They have a hyperbaric chamber there to oxygenate her body to try to keep this infection from spreading." Dr. Romanowski eyed Keith with unshakable conviction. "I sent a woman there for this treatment, and I'm convinced that she is alive today as a result. We will call Pittsburgh and make the arrangements to have them send their helicopter to pick up Isabella, if that's what you want."

Keith eyed Dr. Romanowski, measuring the ardor of his statement. He was certain Dr. Romanowski was attempting to impart an arcane wisdom. This was a man who seemed prepared to stake his professional reputation on the success or failure of a single treatment. It was Copernicus remaining steadfast to his theory that the sun was the center of our solar system, while everyone else insisted the Earth was the center of the universe. It was John the Baptist, a voice crying out in the wilderness, ". . . the kingdom of heaven is at hand!"—Now was Keith smart enough to follow his advice? With a final imploring stare that said, "Do the right thing," Keith followed the faith that was gathering within him.

"If Dr. Romanowski feels that strongly about this, I think we should send Isabella to Pittsburgh." Keith said. He then turned his attention to Stacey. "Do you agree?"

Normally, at times like these, Stacey would prod for more information. She would ask hundreds of questions and weigh the options before making a decision. But in the whirlwind of the overwhelming situation she found herself in, she didn't have the energy for interrogation. She simply nodded her approval.

"I'll make the arrangements," Dr Wilson concluded.

Keith could feel his own resolve growing as Dr. Wilson laid out the details of the transfer. And in the moment of his spirit being broken by yet another tragedy, an unexpected strengthening occurred. His faith in God had found solid ground in the guise of his refusal to accept Isabella's fate. He believed in God's ability to heal, and he believed in his daughter's will to survive. *She has to live,* he thought. *She has to come home again.*

As they left the conference room and made their way back to the waiting area where family and friends anticipated news, Keith and Stacey encountered a young, dark haired physician.

"I assisted Dr. Romanowski during your daughter's surgery," he explained. "I just want you to know, I have two children approximately your daughter's age. I just called home and asked them to pray for your daughter." The expression on his face portrayed his own inner battle. Mental images of a small child, not unlike his own, being nearly dissected alive was gnawing at his own sense of security. He had probably told his children how much he loved them, promised them a special outing during his time off.

Keith and Stacey could see all of his training, all of the medical jargon that once dominated his thoughts, being melted away by what he had witnessed. His eyes alone told the story of Isabella's wounds in greater, horrific detail than the explanation granted by Dr. Romanowski. And for that brief moment, a bond was formed with a man they would never see again.

"Your daughter will be in our prayers tonight," the young physician said before he averted his heartsick eyes to the ground and walked away.

They returned to the waiting the room and the anxious faces that seemed to stare through them. Keith looked at Isabella's Nanny; her cheeks stained with tears, and remembered an apology he had given her only hours ago as Isabella underwent her surgery. Nanny had been gazing out one of the massive windows of the waiting room, her thoughts dancing with the memories she had amassed during her time babysitting her precious granddaughter. Keith had suddenly become wrought with guilt for drawing Nanny so close with Isabella. By making his mom Isabella's caregiver, he had inadvertently drawn her into the turmoil of his life. He was certain their time spent had transformed Nanny's feelings from grandmother to surrogate parent. And now there was nothing he could do to ease the pain of his mom's broken heart, except to ask for forgiveness. "I'm sorry, Mom," he had said. "I know how much you love her." Nanny never discerned the guilt behind the words.

Stacey began to recount the doctor's explanation of Isabella's condition and the impending transfer to Pittsburgh. Keith picked up the conversation whenever Stacey's voice faded into sobs. She told of the removal of the diseased tissue, her voice squelched with anger at the thought of her daughter being slowly and methodically cut into pieces. The room was silent, except for the sniffling of noses and the sounds of a mother's anguish. Stacey's knowledge of medical procedures and their risks was preventing her from viewing her daughter's condition with optimism. Her little girl was sick, and Stacey alone bore the burden of knowing her child's chance of recovery.

Following the harangue, a discomforting stillness fell over the room as everyone tried to digest what they had learned. An air of helplessness settled upon those who came to support the couple through their tribulation. Despite the vast terminology and probabilities, there were many questions left unanswered. The family was in need of direction and Keith was in need of support. After weighing the words in his mind, Keith spoke, "She's going to need lots of positive reinforcement to get through this." With everyone's attention, he continued. "I don't care if I have to quit work and stay home with her full-time. She is coming home. If she loses her leg, we will deal with it." He looked to Stacey for support, but her head was buried in her hands. "She has to come home again," he ended with conviction. His words drew some from feelings of unproductive pity and granted them a call to action. They could help by simply applying faith. If Keith was going to stand against the loss of his child, then they would fight at his side. The power of positive thinking, the marvel of modern medicine and the power of prayer would battle the infection for the life of the child. And everyone accepting the call knew their place in the struggle.

"If you need anything, please call." The statement was repeated with each family member as they departed, giving Stacey and Keith their well wishes and prayers.

Keith could not tell how much time had transpired until the helicopter arrived from Pittsburgh. A young Asian man presented a warm smile as he introduced himself as one of the doctors who would be accompanying Isabella on her flight. The man shook Keith's hand and assured him they would take good care of their daughter. And with that, Keith and Stacey were left to follow by land. Exhausted from lack of sleep, they gratefully accepted Keith's sister, Ann and her husband, Curt's, offer to drive them back to Williamsport. Once back home, Stacey's brother-in-law, Don Brown, had graciously arranged for a car to drive them the nearly four hour journey to Pittsburgh.

They returned home and packed a few days clothing and the necessary toiletries. Stacey's mother, father and sister arrived to see them off and to watch over their house while they were gone.

When the car arrived, Keith and Stacey said their goodbyes and left Williamsport, uncertain when they would return.

In the backseat of the luxury sedan where festive occasions lured revelrous passengers, Keith gazed grimly at the passing landscape. This car had chauffeured teenage couples to their prom, whisked newlyweds off to their honeymoons—passengers with so many tomorrow's they had no reason to look beyond the night. But on this evening, it was embarking on a very solemn journey to a future which, to the marshy-eyed occupants in the rear seat, held little promise for tomorrow.

February 8, 2005

Kathryn Fulmer, known both personally and professionally as Katie, nestled her head back into her pillow and drew the comforter tight under her arms. She was in her late twenties with a fair complexion, wheat hair and a contagious smile. She sighed and ran her fingertips across the cover of the book resting on the bed beside her. The raised print of the title added texture to the word she read *Complications*. The dark red ink reflected the lamplight as though the letters had been crafted from holiday foil. The subtitle, in bold brackets, appended—A Surgeon's Notes on an Imperfect Science.

The author, Dr. Atul Gawande, had delivered a speech at the Carnegie Museum the day before, a speech Katie was unable to attend. Luckily, a friend and colleague of her husband had been gracious enough to secure an extra autographed copy of the book in her absence.

Her bed was a little colder without John, but his absence was something Katie knew would be routine when they exchanged vows. John was a critical care pediatrician while she was employed as a general pediatrician. Thus, their martial abode was often a single dwelling. Working in the same hospital, however, helped the couple maintain some form of contact during their daily routines.

John Polwitoon had completed his undergraduate study and medical school at University of Virginia as a member of the U.S. Air force. From there, he started his residency in San Antonio at the Wilford Hall Medical Center on Lackland Air force Base. Katie did her undergrad at New York University and her medical schooling at Wake Forest. They were both drawn to CHP for the same reason. It was one of the top teaching schools in the nation. At that time, John was two years ahead of Katie in their training, he being a first-year fellow in Pediatric Critical Care while she was a second year resident in general pediatrics. Katie fell in step as a resident practicing what she had learned in school under the guidance of the fellows and seniors. John was introduced to her as one of the fellows who would aid in her training. They fell in love, married and settled into a residential area east of downtown Pittsburgh known as Point Breeze. The broad streets and spacious colonial homes, immortalized in Annie Dillard's *An American Childhood,* boasted the Henry Clay Frick mansion as its most prominent structure. Nearby, Frick Park provides a respite from the bustle of city life for the residents of Point Breeze, many of whom were professionals at Pittsburgh's various educational institutions and health care facilities.

On this night, John had been on call as part of his sub-specialty training, acting as the critical care physician for children who were being life-flighted by helicopter into CHP. He had received an alert call in the waning afternoon hours regarding a possible transfer. That transfer had come to fruition.

Katie picked up the book and quickly devoured the introduction. The writing was seamless and the content; enthralling and easily digested. She paused when she came upon the first chapter entitled, *Education of a Knife*. Indeed. On the job training was a constant in medicine. Knowledge was ever increasing and anomalies forever presenting themselves. For each procedure that was perfected, a dozen more waited in the wings. This was the joy and the gibe of her profession.

Outside, twilight fell cold—a growing nip prevailed inside her bedroom despite the thermostat's attempt to keep the temperature regulated. The yearning for the warming touch of John's body drew her focus. She brushed it aside. The book would have to serve as a surrogate mate for the evening. She would be fast asleep by the time he made his return.

"Are you awake?" John's question had barely passed his lips before Katie uttered her response.

"What's up?" Years of regular call nights had manifested both behavior and biological changes in Katie. Waking up groggy from a deep sleep was a thing of the past. Rousing was now reflexive, and surprisingly clear-headed. The nature of her job showed in her inherent response. What's up? What's wrong? What do I need to do to help?

John's face, eidetically crescent-lit by the runner of light coming from the gapped door of the master bathroom, was flexed with angst. He removed his glasses and rubbed his eyes; his gentle features had shifted from calm to disquiet. "We transferred a two-year-old girl with necrotizing fasciitis from Geisinger Medical Center. They performed a debridement, but the infection is still active." He sighed. "God, you should have seen it. I literally watched it spread during the course of the flight."

Katie studied his expression. The experience had visibly shaken him. As a physician, John's most impressive trait was his tranquil demeanor in the face of emergencies. She had watched him closely in the beginning of her training at CHP. He was unflappable. And she tried to learn from the example he set. No emotions on the surface. Focus on the task at hand and you will be a better doctor as a result. But this, this was a side of him she rarely witnessed.

"She's being scheduled for hyperbaric." He slumped down on the edge of the bed and pulled his socks off.

"Do you think it will work?"

"I don't know. This infection's impressive." In physician argot, impressive was not a term of admiration for a disease, but an expression of dread. "I don't think we can stop it. I think we're going to lose her."

Katie slid across the bed, pressing her chest against his back and wrapped her arms around him. Her head nestled itself between his shoulders.

John leaned back into her embrace and fell silent. Dr. John Polwitoon was only a part of the man—a mask of confident, mechanized healing which fought to conceal any fear of failure. But tonight, his paternal instincts had cowered to the suffering of a tiny girl. And in the refuge of his home, Dr. Polwitoon allowed himself that momentary weakness.

Pittsburgh, Pennsylvania

The car rested outside the emergency room entrance of the Children's Hospital of Pittsburgh. It was 3:00 am. The streets of Oakland, a suburb and the cultural epicenter of Pittsburgh, were nearly deserted in the thick of the winter evening. The cold air stripped the heat from Keith and Stacey's outer layer of garments as they climbed from the car and stared up at the concrete edifice of the campus as it rose and faded in the blackened sky. The structure of the hospital was slightly over ten stories tall and the surrounding buildings resembled a college campus.

The driver removed the couple's luggage and, following an awkward moment, accepted a tip from Keith, who was unsure of the etiquette in such an unusual situation. The driver seemed appreciative, offering the couple his best wishes before departing to the east.

Under any other circumstances, the trip would have been pleasant, but the warm and comfortable car did little to ease their tension. They tried to catch up on some much needed sleep, to no avail. At one point during the trip, the first doctor to examine Isabella upon her arrival called Keith on his cell phone to let him know she had arrived safely and was stable.

The security guard inside the automatic doors gave the couple directions to the Pediatric Intensive Care Unit, or PICU. They loaded their bags onto a makeshift cart and headed for the elevator. Keith thought the hallways appeared more like tunnels, far more visually suited for the lower levels of the Overlook Hotel in the movie *The Shining*. The passageways were squat with walls painted pale beige. The ceiling was equipped with sparse, recessed lights which emanated an ominous and dreary atmosphere. Water pipes ran audaciously exposed just below the wall and ceiling joint, and the multi-colored tile floor was dingy. Keith could almost hear the voice of some, fresh out of the freezer, axe welding psychopath shouting the maniacal self-introduction of "Heeeeere's Johnny!" Or perhaps it was just Keith's perception as filtered through his current mood.

The elevator stopped on the sixth floor and opened up to a slightly happier atrium complete with an eight paneled light fixture designed as a counterfeit skylight. They made their way to a large set of double doors leading to the PICU, paying little attention to instructional signs or directional lines on the floor. They walked up to the doors, but the doors didn't open. As they pushed and pulled trying to gain access, an agitated female voice came from a speaker mounted to the wall.

"You have to step back behind the red line or the doors won't open," she scoffed.

How can someone use that tone of voice to people in our situation, Keith thought. *No matter how many times she has to repeat herself to heavyhearted and weary parents, she needs to remember where she works.* It was like the doorman from the Land of Oz shouting at Dorothy and her group of misfits for ringing a doorbell that apparently worked, but was out of order. "Didn't you read the notice—it's as plain as the nose on my face." In this case, it was a door that worked only if you stood behind a red line. Here, there was no yellow brick road; no magical Land of Oz; no ruby slippers to appease the troll on the other side of the door; and no wonderful wizard. *Too bad, this rude woman is in dire need of a heart.*

They stepped back behind the red line and the doors magically opened. As they passed through, they were greeted by the Grumpy Door Troll who abruptly explained the PICU's ground rules for parents and visitors. They would have to check in at the front desk each morning to receive a PICU pass. In addition, they would have to remember a four-digit code to access patient information. They would be permitted to sleep in the waiting rooms, although the hospital did reserve a dozen 'sleep rooms' which were granted via a lottery system.

"Your odds of getting a 'sleep room' will depend on the number of parents that are staying over on any particular night," the Grumpy Door Troll explained. "You will receive a packet tomorrow when you check in that will detail all of this and more. Isabella is in surgery with Dr. David Becker. You folks can sit in one of the waiting rooms and he will be in to speak with you as soon as he's finished."

Patience never comes easily when you're trapped in the alternate universe of dread. Keith and Stacey sat in one of the vinyl arm chairs until their restless legs prompted them to pace the small confines of the waiting room. It was as though they had been caught up in a tornado, twisting about on an insane ride and then dropped into the middle of a foreign city, now awaiting the arrival of yet another physician, Dr. Becker. At last, he appeared—he was young, dark-haired and bespectacled, and donned the customary scrubs of a doctor returning from surgery.

"Hi, I'm Dr. Becker." Keith recognized the voice as the doctor he had spoken to on his cell phone during his journey to Pittsburgh. "I am one of the surgeons here in the children's hospital. Isabella is very critical and the next twenty-four to forty-eight hours will tell us a lot about her chance of survival. I took a look at the work that was done by Dr. Romanowski in Danville, and I was impressed. He was very aggressive and with this type of infection, that is exactly what is needed. I had to remove a little more dead tissue from her back and leg, but it looks as though he got most of it. There may be more debriding over the next few days or even weeks, depending upon how quickly we can get a handle on the infection. Tissue that is dying now will show itself later and will have to be removed."

"What about her leg," Stacey asked.

"Well, she's lost a great deal of muscle. I really don't know if we can save the leg. We will have the plastic surgeons look at it tomorrow as well as my partner, Dr. Selvey. It's just too soon to tell. The main thing to remember is that we have to save her life first and foremost. If she has to lose her leg, we will cross that bridge when we come to it. But we'll do everything we can to avoid that as we get her though the next forty-eight hours."

"Can we see her?" Keith asked.

"Of course," Dr. Becker said, "as soon as the nurse gets her settled in, she'll call out and let you know."

Stacey held her comment until Dr. Becker had disappeared, "I can see where this is going."

"What do you mean?" Keith asked.

"They cut and they don't care. They'll keep her alive no matter how much of her they have to cut away. And they'll never consider her suffering and pain or what the quality of her life will be."

Keith had always preferred his healthcare served with a topping of faith, not caring to see what was underneath. Now a popular movie came to mind. A group of space travelers became marooned on a planet inhabited by nocturnal, carnivorous alien-creatures that enjoyed eating people's heads. Lucky for the astral-castaways, the predators couldn't tolerate any light source, and this planet revolved around three suns. But as things go, along came a neighboring planet that looked reminiscent of Saturn with a jonesing for a dramatic plot turn and an orbit which would place it between their new home and the suns—bringing about a total eclipse. You can never trust a planet with rings. As darkness prevailed, the makeshift band of humans tethered themselves together with a battery-powered glow-hose that looked like it had been borrowed from a Rave and attempted to traverse the alien's terrain to safety before they became a midnight snack. At one point, one of the survivors, a wimpy antique dealer, became so terrified that he tried to crawl away, disconnecting the glow-hose from the power source. In utter darkness, he used a mouthful of liquor and a lighter to turn himself into a human blowtorch, in a kind of Kiss meets the phantoms of the planet. When the flame erupted, he realized he was in the middle of an evil alien huddle.

In the case of Isabella's medical care, Stacey had just spit that illuminating fireball and Keith wished he had not seen the true gravity of the situation. Although not knowing they are there, doesn't make the aliens go away. Isabella's tiny body had been invaded by microscopic aliens and her only chance of survival was in knowing the bacteria were present and the knowledge of how to eradicate them.

Once again, Keith and Stacey found themselves in front of the doors leading to the PICU, making sure they remained behind the red line. The Grumpy Door Troll depressed the button from within, a buzzer sounded and the doors opened. She led them back through a gut-wrenching maze of occupied beds. Babies and toddlers were lying in the dark with machines pinging and buzzing as a labyrinth of tubes and lines pierced their young flesh or obstructed orifices. Infants clinging to life only by means of respirators and feeding tubes, while heart monitors and BP machines kept sentinel over the status of their condition, were separated from one another by thick curtains. The line of beds along the far walls to their left seemed endless, spanning three sides of the room's perimeter. The fourth side was occupied by a nurse's station complete with computer screens wired to each patient's monitors. They continued their way around the gauntlet of suffering, passing by every bed in the expansive room and baring witness to the desperate struggle of each of the tiny occupants. The sadness of the room was overwhelming. *Isabella doesn't belong here,* Keith thought. *No child belongs here.*

In the wee hours of the morning, it was more than understandable that the parents would not be present, yet their absence added to the eerie display. It was as though the children were left here, abandoned to bear the cross of their mortality without the aid of a nurturing caress or gentle words of encouragement. *These children should be home sleeping, perhaps even with their parents snuggled up next to them; waking to brothers and sisters who act as playmates; filling their little stomachs with eggs and pancakes or cereal; spending their day without pain.*

The close-up look at the PICU burnt an indelible scar into their memories and made the world appear a lot uglier.

They rounded the final corner where the line of beds gave way to isolation rooms that were reserved for the sickest of the sick. These were 'clean' rooms which hosted those who were contagious or immune compromised. Isabella's room had a windowed door to the left and a large window to the right where a small nurse's station monitored the patient from the hallway. The door led to an ante room with a sink and cabinets above and below which stored emergency medical necessities along with fresh gowns and surgical masks. Above the waist, the walls that separated the room from the hallway were large thick glass encapsulated by metal frames which provided a panoramic view of the patient's room. Large Venetian blinds stretched across each windowed wall to protect the loved ones from seeing procedures they would find discomforting. For now, these blinds were opened. From the ante room, visitors entered the patient's room via an adjacent door on the opposite side.

After thoroughly scrubbing their hands and attiring themselves with gowns and masks, Keith and Stacey entered Isabella's room. They found her lying motionless on a railed bed with lines, wires and tubes invading her tiny body in every way possible. Before departing from Geisinger Medical Center, she had been intubated and placed on a respirator, the plastic flex tube now protruded from her mouth. IV lines extended from both arms, a subclavian line was inserted in her shoulder and a Foley catheter in her bladder. Her medications were numerous including antibiotics of Piperacillin, Tazobactam, Clindamycin, Ciprofloxacin and Fluconazole; Epinephrine titrated to maintain blood pressure; pain medications of Fentanyl and Versed; transfusions of platelets, packed red blood cells, albumin and fresh frozen plasma. The skin on her face was puffy and stretched tight from edema which was the result of the many intravenous fluids and steroids that were being pumped into her tiny veins. Monitors of all shapes and sizes flanked her bed and an eerie silence hung among the background of the all-too-familiar sounds of subtle clicks and beeps. Her chest rose and fell with impetus of the iron lung. The walls inside the room were painted the same bland beige which only served to make the room appear smaller and more maligned—an ambience far removed from the playful pastels of her bedroom at home. The only remaining link to her coddled life at home was the ever diligent *Blankie*; still close at her side. Yet here, the entire chamber cried *hospital* and all its tones played a song of frailty.

A middle-aged nurse with furls of chestnut hair had just finished an injection into Isabella's IV. She offered Keith and Stacey a warm, condoling smile, "Hi. I just gave her a pain shot. We try to watch any reaction to pain; any irritable movements or increases in blood pressure and heart rate. We're trying to keep her as comfortable as possible."

"I don't want her feeling any pain." Stacey said sharply.

"We're going to do our best." The nurse gathered the remnants of her instruments and eased from the room.

Isabella was quiet, motionless. But within the still facade of her body, who could say what coursed through her nerves. The battle for their child's life was all but out of their hands, nothing to do now, except wait. Time was theirs to think, to cogitate what was, what could have been and what will be; leisure for deals with God and musings of one's unworthiness of redemption or favor; time to test the endurance of one's faith. The waiting had only just begun.

"Forward this email to ten people and a miracle will happen."
So read the instructions at the bottom of the email on his computer screen.

It was the typical carrot given IF you follow the instructions, which, of course, would forego the customary threats of bad luck if you don't. The only question beyond the decision to send it was who starts these things? Control freaks and terror mongers who enjoy intimidating people into following their directions? Or is it nothing more than an ego trip, the satisfaction in knowing that something you created is being transferred all around the world many times over. These bites of pop-culture only meant to stroke the creator's pathetic sense of accomplishment, a proud moment in their parent's life. "My son created an email that is still being passed around the world after two years!" If only by the chronically gullible.

Gullible was not a word used to describe the email's recipient, Don Brown. Don was a self-made man almost from the moment he stepped out of high school and an entrepreneur before production ever began on his first invention, an abdominal workout machine.

He was an optimist, but not the naive Hermey-The-Dentist-Elf type who would run away with the first reindeer outcast with a phosphorescence problem. He was closer brethren to Yukon Cornelius; guardedly optimistic. But something on this particular day, at this particular time, had him ready to open his address book and throw his proverbial internet hat into the ring.

His niece, the only child born to his sister-in-law, was apparently in the throes of death. He knew the girl was sick. And he knew she had been taken to an emergency room in Pennsylvania with a flesh eating bacteria. But now his wife, Tanga, who was at her sister's side, had given word that Isabella would not make it through the night. Earlier, his wife had called to inform him that he should probably pack a dark, formal suit for the impending funeral. He then called his brother-in-law to express his sympathies, "You don't deserve this," he had told Keith, who responded with a solemn, "I'm not leaving here without my daughter."

Positive energy was a good thing. Even in the face of such insurmountable odds. And during these helpless situations when physical response fails us, our faith instinctively takes the wheel, driven by our own belief in forces beyond our skin. The deep primitive well within each of us will produce a drop of credence that we pray will invoke a flood of optimistic water to cleanse the evil that surrounds us. So that in the dew licked morning after, the sun can shine hope again.

In the world of blind faith, even an email chain from an ego-driven, fear-mongering, author of junk internet mail can present itself as the first link in a chain of events to rectify any conundrum. Don suddenly felt compelled to comply with the demands for a miracle; to forward the email, but not because he was bowing to some irrational superstitious behavior. He would do it as a display of positive thinking and in doing so; this request for a miracle would be sent indirectly to God via the otherworldly wide web. Don opened up his address book. Soon, ten of his family and friends would be faced with their own allegorical dilemma.

February 9, 2005

The rising sun did not signify the beginning of a new day, only a lighting change during a one act play more tragic than any string of Shakespearean soliloquies. The morning brought new characters discussing more surgical procedures. A parade of doctors like soldiers of pessimism who marched in step repeated, "We had to cut away more tissue, and your child may not survive." One doctor told them it would be a miracle if a child this sick did not develop some kind of organ failure. In the meantime the veracious bacteria continued their unrelenting feast, eating away at her muscles. Her back and her right leg were still slowly dying.

Keith and Stacey found themselves back in the cramped quarters of the nine-by-twelve foot waiting room while Isabella underwent another round of surgery. Once again, the couple held their collective breath and prayed like never before that the debridement would cease.

It was early morning, though the PICU floor lacked the windows to see the new day break. Not long after Isabella was taken to the OR, they were visited by a Dr. Marian Hauser from Infectious Disease Services who began running through a prescript line of questions which started out innocently enough. Is there any family history of immune disorders? Any family history of cancer such as leukemia? Any recent abrasions; open wounds; skin rashes; abscesses; boils? What types of infections had Isabella had up to this point? But then came the unexpected.

"Is there any possibility that the two of you are related?"

Keith and Stacey's heads swiveled toward one another as though they were compass needles drawn to intersecting polarities. The look in their eyes bore a hodgepodge of unease, curiosity, and the affirmation of the absurd. The line of inquiries had suddenly narrowed to bizarre speculation more suited for afternoon television. Meet Stacey, she thought she had fallen in love and had a child with the man of her dreams; instead she found the cousin she never knew existed.

"Believe me," Dr. Hauser continued, "this question is as strange to ask as it is to answer. But it's something we have to rule out."

"There's no way," Keith finally said.

"This morning," Dr. Hauser continued. "We will be taking some wound cultures from both the back and groin areas during the surgery. The cultures will help us narrow down the exact strain of bacteria with which we're dealing. We should have the results in the next twenty-four hours."

Shortly after Dr. Hauser departed, they received word that Isabella had returned to her room following her surgery and recovery. Stacey had gathered her emotions. Motherhood has a way of granting a woman the ability to quell her fears with a tenacious desire to protect her child. It is a perfect balance of nurturer and defender. She could fall apart emotionally later, now she marched with deliberate intent through the labyrinth of pediatric despair to visit her daughter. Keith trailed his eyes ever aware of the suffering which surrounded him. He had chosen to wear his wrought on his sleeve. Unashamed of the sorrow he felt, though his open display of grief was not without purpose. If the doctors could see how much he loved Isabella; how desperate he was for her recovery they would be more vigilant. They would try harder to evoke a cure.

In a chance consultation, Keith and Stacey had once again entered the ante room to sanitize their hands and dress themselves in germ barrier gowns and masks before continuing into Isabella's room. Dr. Edward Selvey, a stocky, bespectacled black man, had just completed his post-surgical examination and was just exiting.

A respected member of the medical community, Dr. Selvey's had passed through the halls of Yale University and Harvard Medical School on his way to acquiring his positions as both pediatric surgeon for the Children's Hospital of Pittsburgh and associate professor of surgery at The University of Pittsburgh School of Medicine. It was comforting to know his skills had been utilized to treat Isabella, but it was disquieting to hear his opinion, the same grim outlook. A nurse who had assisted him in the exam was at his side; her eyes looked upon the couple with sympathy—Dr. Selvey's eyes were phlegmatic to the point of mechanical. It was a look that Stacey found all too familiar.

"We cut away a little more tissue, but it's still too early to tell. There is a possibility we will have to remove her leg," Dr. Selvey explained. And with each word, a fresh flow of tears streamed down Keith's cheeks. His icy blue eyes deepened with the shimmer of fear, adding an almost childlike quality to his features. His lids were torrid, inflamed fault-lines with puffy, ashen sacks contrasted by the pale skin surrounding them. He postured himself in submissiveness; shoulders slumped, head lowered. If necrotizing facititous could somehow form itself into a human figure he would have gladly, and quite violently, torn it apart. But this was a creature he didn't know how to fight and in that helplessness, his body language spoke volumes.

Unaffected, Selvey droned on. "She had extensive muscle necrosis on her back, thigh, and groin areas. On her back the paraspinous muscle and latissimus dorsi capsules were opened and excised down to the ribs. Only her intercostal muscles remain in that area. In her right thigh area, the anterior medial compartments of the femoral muscles were debrided down to just a thin layer which covers the bone." He removed a small device from his pocket, "I have images of her wounds if you would like to see them."

Stacey felt a quiver course the length of her spine and she nervously ran her fingernails through a tangle of black hair, which was once well groomed. She vehemently shook her head. She feared seeing the pictures of the wounds; feared the raw images would break her. Feared she would no longer be prepared to make the difficult decision of letting them proceed with the surgeries. The wounds would make Isabella's suffering all the more real and she would demand they stop cutting her. Keith was too stunned by the question to produce a response.

"If she has to lose her leg; will she be able to sit up on her own?" Stacey locked onto Dr. Selvey's eyes searching for a glimmer of the human that lied beneath the calculated demeanor of the surgeon. "How far will you go before you say enough is enough?"

Dr. Selvey was visibly startled by the question. It was an irregular response that didn't fit into the typical 'doctor-parent' protocol. "Despite any physical abnormalities, it will still be Isabella."

"Let me ask you . . ." Keith paused as a wrenching sob surfaced in his voice. He choked back the tide of emotion with a clearing of his throat. "Let me ask you a question? Do you have children, Dr. Selvey?" Keith asked.

"Yes. I do."

"Don't ask that," Stacey snapped. "He's not going to answer a question like that."

"What you're doing to save Isabella," This time the sob invoked a Richter-worthy quiver from his body, prompting a lightly placed palm from Dr. Selvey's accompanying nurse upon Keith's back. The gesture seemed reflexive as she withdrew it as quickly as she placed it. "Can you look me in the eye," his voice broke off, but he incessantly pressed on, "and tell me that you would do it to your own child?"

Dr. Selvey's lips tightened, "I've had to do a lot of things to children that I wouldn't do to my own child." He made his way around the couple within the confines of the small room and slipped out the door. The softhearted nurse followed in his wake.

Stacey swallowed hard against the lump swelling in her throat, not born upon sorrow, but upon disgust. "I've seen this over and over—they will just keep cutting away. They're only concerned about saving lives no matter what the consequences."

Keith allowed the thought to roll over in his mind. *Does the value of human life diminish with the quality of life?* The question was far from rhetorical. It was a beast with its claws latched firmly around his heart and its fangs sunk deep into his throat. In the passing of forty-eight hours, it was becoming the only thing that was real, as all insignificant thoughts of appointments, schedules, and what to cook for supper were brushed by the wayside. *How far would they have to go? No matter what, it's still Isabella.*

"I'm not ready to give up," he said, as he wiped the back of his sleeve across his face then slipped a gown over his clothes.

"Neither am I," she responded as she tied a surgical mask behind her head. "But if it goes too far, we are going to have to be the ones to speak up for Isabella."

Keith shook his head. "Even if we were to reach that point, do we have the legal right to make that kind of decision?" *Who was the woman in the news? Schiavo? Yeah. Terry Schiavo. Her family is trying to keep her alive while her estranged husband wants her removed from life support. Her family refuses to give up on her, but it appears destined for the court system.*

Stacey entered Isabella's room, allowing the door to close behind her. Keith finished securing his mask as he watched through the glass at Stacey approaching Isabella's bedside. Stacey took her position on the side of the bed most accessible to Isabella and held her hand.

Isabella's leg shifted. Stacey turned to motion for the nurse through the large glass observation window as Keith arrived at the bedside. "She's in pain," Stacey announced.

Later that night, the couple was lucky enough to win one of the PICU 'sleep' rooms which consisted of narrow, confined chambers, approximately eight feet wide and twelve feet deep with murals painted on the walls which appeared to have been decorated by a burger clown's evil twin on an eight day acid binge. The rooms were granted to the exhausted family of sick children for much needed rest and privacy. The remaining families were forced to find refuge in waiting rooms and quiet rooms.

Unfortunately, this particular sleep room did not live up to its name. Just beyond the wall where the headboard nestled, the elevator continued its labor throughout the night.

Lying on the bed with his back pressed against a rigid mattress and Stacey curled on her side next to him, Keith was exhausted but unable to sleep. Night had fallen and among the thousands of people who occupied the building around him; he was alone. His battle-numbed mind wandered to Isabella lying amidst all of the suffering; the tubes and wires monitoring her condition, some of the machines even keeping her alive.

His eyes had adapted to the low light enough for him to make out the dark edges of the angular brush strokes of the wall mural which leapt from the murk like images plucked from Pablo Picasso's most perturbed nightmares. The silence within the room was intermittently broken with the monotonous hum of the electric motor, the tension pops of cables and the 'ding' as the elevator perkily announced, once again, it had achieved its destination. *This is a godforsaken place,* he thought.

Nanny's mind leapt from image to image, flashing like a strobe light out of control. Each passing smile; each happy or loving memory was now met with pain, the physical manifestation of remorse. A suffering she recognized as pain, yet it had no point of origin. She knew she was agonizing, but she could not point to where it hurts. A lifelong battle with an undiagnosed nerve condition had always surfaced at times like these. Crying was the easy part. Her body shook uncontrollably and her lips tingled with numbness. Her inner organs felt as though they were being mixed in a blender.

The first fifty years of her life were far from blissful, but she rarely felt the sting of loss until the death of her granddaughter, Samantha, set off a chain of events in which each tragedy seemed to fade into the next. One year after losing Samantha, her husband died; the following year an attempt was made on the life of her eldest son, Ralph, which left him in a medically induced coma for three months while his body recovered from third-degree burns and damaged limbs. During the first month of Ralph's hospitalization, she was involved in a car accident with her son, Glen, and her daughter, Cindy, while returning home from a visit. Ralph was a patient at the Lehigh Valley Hospital burn center, some one hundred and thirty miles east of Williamsport. The trip took them across a section of Interstate 80, a transcontinental corridor that connected San Francisco to the Hudson River in New Jersey through 11 states. It was one of the most traveled thoroughfares in Pennsylvania, smirched with countless traffic accidents. It had rained for most of the afternoon during their visit and the cold highway had become slick. At one point during their return trip, they stopped at a fast-food restaurant and grabbed some burgers and French fries to eat on the way. Glen drove with Cindy in the back and Nanny sitting in the front passenger seat. The tension since leaving the hospital had eased and Nanny was content as she focused on dipping her fries into barbeque sauce before eating them.

"Mama," Glen growled, as his knuckles whitened from his grip on the steering wheel. "I don't know what to do. I can't stop, we're going to hit."

Nanny responded with little thought, never looking up to see the view through the windshield as the rear of the 18 wheel-truck in front of them grew ever closer. "Oh, you'll be okay. You can do it." She happily soaked a fry in her condiment and drew it up to her mouth.

"This is going to hurt," were Glen's final words of warning.

When she regained consciousness, she was wrapped in a neck-brace and strapped to a stretcher as paramedics carried her to an awaiting ambulance. Her head had hit the windshield and, from the lack of a seatbelt, her body had bounced off the dashboard. Glen and Cindy managed to brace themselves and escaped with minor abrasions. She was taken to Geisinger Medical Center where she was x-rayed and released with no extensive damage. A few days later a persistent pain in her chest sent her to the Williamsport ER were she discovered she had fractured three of her ribs.

In the second month of Ralph's recovery, her mother died. But with Ralph's triumphant return home, a five year calm settled into her life until the death of her father marked the beginning of another reign of tragedy. Her brother, Richard, died in July of the following year and her only sister, Betty, was delivered to God in November. Before she could catch her breath, her beloved brother, Sonny, was diagnosed with leukemia.

And now the only person to help her recover and break the chain of despair; the linchpin who had lifted her from sullen to joyful, was now in another town struggling to survive. Pittsburgh may as well have been Tibet for a woman who had done little traveling in her life. Isabella was miles away.

Keith had telephoned her earlier to inform her that Isabella's latest surgery went as well as could be expected. The doctor's were cautious in predicting her premature recovery, but, for now, Isabella was still fighting back against the deadly infection.

Nanny sipped at a glass of diet soda, allowing her eyes to scan the many toys organized in displays around her living room; Isabella's signature on her life. The quivering of her lips foretold the return of tears before they ever reached her cheeks. It was as though each toy took its turn by standing up and reminding her, "Isabella's not here." A swirling memory settled like a drift of snow near a silent playground. And once again, she was holding Isabella tight in her arms singing, "You Are My Sunshine." The tension was stretching her, testing the elasticity of both her soul and her skin. If she did not change something, soon, she was certain she would be ripped asunder. Her crime was caring and her punishment was to be emotionally drawn and quartered. And in her frail state of mind, it seemed the toys were the horses that would draw the ropes taught. She needed to distract herself from the pain; even if it were only a brief moment of detachment.

Nanny had always been the consummate mother and housewife. Through the best of times, she cleaned. Through the worst of times, she cleaned. To keep herself active, she cleaned. To keep herself calm, she cleaned. The idea occurred to her; a single answer could provide healing in duality. She could busy herself by moving Isabella's toys into the spare bedroom on the second floor, thus removing the reminder of Isabella's absence.

Please don't take my sunshine away. The lyric entered her mind and she knew the only way to hold herself together was immediate action. She rose from her seat and gathered an armful of brilliant colored plastics, molded into happy shapes and furry mocks of animals, stuffed with plush materials until their seams bulged. And with her treasure of playful bounty, Captain Feathersword's first mate shuffled sadly upstairs.

It was a morning ritual. An uninterrupted drift of pedestrians made their way through the streets surrounding the University of Pittsburgh Medical Center and the affiliated Children's Hospital of Pittsburgh. Nearly all of them coming or going from the hospital, doctors, nurses, students, patients and families of patients moving along the perimeter of the facility like drifts along a snow bank. These days, the flow of traffic along Fifth Avenue was compressed down to a single lane while construction lumbered forward on a new building being added to the already immense teaching hospital. In the early morning hours, most of the foot traffic consisted of doctors seeking a grab and go breakfast at Bruegger's or Panera. If you were a member of a team, the pecking order of physicians was obvious; it was the neophytes who were sent forth to gather the grub and coffee.

Victoria Street to the north as well as Fifth Avenue to the south was the main artery with the dissecting Darragh, Lothrop and Desoto Streets acting as the capillaries. Victor Street also served both as a drop off for UPMC's emergency entrance and the northern entrance of CHP. This morning, it was also being utilized by a team of doctors and nurses as a passageway from one hospital to another.

Together the Stryker bed with Kinair floatation mattress, designed to provide optimum comfort for patients with compromised outer tissue, weighed nearly 700 pounds. At seven feet long and nearly four feet wide, it was over a quarter of a ton of hard-to-maneuver, kinetic energy on wheels. The tiny passenger, poised on top, added very little mass to the behemoth bed, yet there were seven staff members assigned to the transfer. Their task was to cross Victoria Street wheeling the Stryker which carried a sick toddler with numerous IV lines. Earlier, they had gathered all of the essentials for the jaunt. Pumps, transportable poles, additional meds (some drawn only in case of an emergency), a transport monitor, all had to be collected and put into place. Her ventilator and IV tubes had to be properly positioned so they were long enough to reach when she was placed inside the hyperbaric chamber. They had to remove her from the ventilator and hand bag her for the trip to where the chamber awaited in UPMC. Then they moved the bed from the isolation room, through the PICU to the elevator and down to the hallway on one of the lower levels.

Dr. Fulmer and her colleagues eased back as the weight of the bed now powered its own propulsion. This was an easy stretch of hallway. The difficult part would be on the open street.

The patient was Isabella Cole. Fulmer was already aware of Isabella's presence in the hospital. John had taken a keen interest in the child's case during the transfer to CHP. Fulmer, herself, had also examined the young girl—made physical contact with the patient, something she did routinely. Placing her hands on her patients made them more real. They were no longer symptoms, prescriptions and stats; they were people in distress who were in need of her help. They had pain and suffering; they had loved ones who were concerned for their well-being.

Only a percentage of healing could be taught in a book. It was the doctor-patient experience which engendered the most knowledge of the body and its intricate design. And in the form of this little girl, there was much knowledge to be gained.

The doors loomed ahead.

They began to brace themselves, pulling back to ease the bed's progress. A rush of cold air greeted them as they made their way through the egress and into the open street.

A small section of Victoria Street was a porte cochere as it passed directly through the conjoined buildings of UPMC and CHP. It was used by vehicles accessing both the hospital entrances and for thruway traffic. Along the sidewalks, hospital staffers passed from building to building, visitors could be seen often times moving with purpose and occasionally meandering like lost puppies. Ambulances sped up to the ER entrance of UPMC on nearly regular intervals and unwitting motorists passed herky-jerky through this particular section of Victoria Street, wishing they had chosen an alternate route.

Near the buildings, smokers huddled together against the cold to soothe their nicotine urges. The second hand smoke rose up and became trapped by the overpass before drifting back down in a swirling pother above the scuttle of activity. Fulmer grimaced at the inauspicious cloud. She would almost bet a week's wages that at least one of the smokers was here as a result of someone having lung cancer. Fortunately, Isabella was being ventilated.

Isabella's entourage was forced to bring the awkward moving bed to a complete halt while a few passing cars made their way along the tar macadam. After the all-clear was given, they willed the bed into motion, pushing it onto the blacktop as they regained their kinetic speed. Attempting to steer the Stryker across the street was like guiding a stubborn bull through a mine field. Fulmer glanced down at the little angel resting all but comfortably on the mattress. Her lips were already beginning to dry; the vent tube protruding from between them was like a thorn through a rose petal.

"Whoa!" A male co-worker at the foot of the bed shouted a warning. An approaching sedan was determined to make his pass despite the clinical caravan.

The staff instantly tightened their grip and locked their knees against the beds perpetual motion. For awhile it seemed a collision was imminent until the bed rolled to a stop.

A fire flashed in Fulmer's eyes; one not often seen. "For God sake, can't you see the condition of this child?" She glared through an angry squint at the rear window of the car as it continued along the road.

After a sigh of relief, the crew pushed the bed forward across the remaining street.

"Some people ignore suffering for their own sanity," the male at the foot of the bed spoke in a disgusted whisper, "some people are apathetic and some people are just pure assholes. He was definitely the latter."

Fulmer made the final unnerving preparations for Isabella's insertion into the chamber. Once the pressurized chamber was sealed, there would be no way of opening it emergently. If her endotracheal tube came out of her airway during pressurization, a chance existed for Isabella to lapse into cardiac arrest before the chamber could be opened.

The hyperbaric chamber consisted of a horizontal cylinder of thick, tempered glass capped on one end with a reinforced steel door which nearly resembled that of an industrial furnace. The base was a rectangular box with a raised control panel and four L-shaped legs bracing it a few feet off the floor.

With everything in place, the chamber was sealed and the treatment began without a glitch. Fulmer settled into a chair and kept a watchful eye on the monitors. Inside the apparatus, Isabella looked like Snow White asleep in the glass casket after biting into the apple given to her by the witch in the guise of a kindly old woman. But in this fairytale the handsome prince was the culmination of countless years of research from the blind dedication of thousands of mankind's most brilliant thinkers. Modern medicine would provide this kiss if God could provide the magic.

Isabella's vitals looked good. Fulmer relaxed a bit. Valentine's Day was just around the corner; she had to come up with some ideas on what to buy John.

February 10, 2005

Like a babe from the womb, delivered into a strange world of eye piercing luminosity, new faces and alliances, words of unintelligible gibberish and indecipherable jargon, frightening machinery and an existence in which you are lead around—carted from place to place with instructions on what to do and when to do it, the only power Keith and Stacey possessed in the first three days was their ability to observe and learn. Most of their energies were focused on ingesting as much about their daughter's condition until the information became physically nauseating. In the periphery, they learned time schedules, where to eat and sleep, how to enter the lottery to win a 'sleep room' for a night, and what was the expected behavior and rules of their new home. They were not alone in this quest. They shared this quasi-domicile with other families of varying experience; families with children who were battling their own diseases and ailments. Each little life had a story balanced on the gossamer thread of a heartstring. The waking hours seemed disjointed as they became acquainted with each family through the fog of nearly catastrophic events.

The day after their arrival, they met a couple named Cindy and Terry, who along with Cindy's parents, took everyone in the waiting area under their wings. Cindy and Terry's son, Ian, was admitted to CHP suffering from a rare reaction to a new anticonvulsant medication which subsequently unleashed a disease known as Stevens—Johnson syndrome. Ian, already battling cerebral palsy, was now forced to endure a painful life-threatening rash which formed in patches on his body and in his eyes and mouth. Similar to third degree burns, his skin began to blister and became loose, easily rubbing away from his body. The disease caused excruciating pain accompanied by chills and fever.

Despite all the reason in the world for self-preoccupation, Cindy and Terry were more in tune with the suffering of others; the resident tour guides for all things good in the human heart. They played the Eucharistic ministers to Cindy's parents, who passed out the sacraments of blankets, food and comfort. Cindy was a kindly extrovert with a loud and boisterous demeanor while Terry carried himself with an understated country temperament, but together in this atmosphere, along with Cindy's parents, they were a shroud of solace for everyone around them.

Less charitable, but just as present, was a candidly divorced couple awaiting word on their seven-year-old daughter who had been recently brought in with a high fever and signs of pneumonia. The woman's estranged husband was joined by the new man in her life, who apparently felt it necessary to vie for the woman's attention despite the condition of her daughter.

A less combative couple had a daughter the same age as Isabella who developed kidney failure. Her body was bloated; and her skin jaundiced. Her features were so distorted in size and color it was nearly difficult to recognize her as a toddler.

Yet another toddler, this one three-years-old, was admitted after her parents found her seizing in her crib with a high fever. Faith, as her parents aptly named her, was battling an infection of her own, dissimilar to Isabella's.

A twelve-year-old boy, wheelchair bound and fed by tubes, was accompanied by his middle-aged father, a bewhiskered man with a limp. This paternal sentinel was isolated, perhaps by choice, but more apparently by fate, with no family or friends to support him. Because of his son's extensive needs, the limping man could not work, so his son was the only person in his life. He spent most of his initial days in the hospital sitting in an armchair in the least populated corner of the room. Until, at last, Cindy's parents eased themselves into his life through simple acts of kindness and managed to open his world to a new set of friends.

The reports Keith and Stacey were receiving on Isabella's condition continued to be bleak. They channeled the information back home to their respective families through single contacts who were then in charge of spreading the information throughout the remaining members. But on this morning, Cindy, the resident waiting-room angel, told Stacey of a program the hospital offered called 'Care Pages.' The hospital's internet server provided each child in their care with their own web page of photos, postings of the patient's progress as written by their parents, and a message board for well-wishers to post their comments and prayers. It was an umbilical, techno-interface of life sustaining communication to and from the parents at the hospital and their families, anxiously awaiting word, at home.

Stacey ascended to the seventh floor of the facility. Just beyond the 'sleep rooms' was the entrance to the hospital's library where she discovered the computer of which Cindy had advised her.

Part of the initial setup was the designation of the password for Isabella's personal web page. For this, Stacey chose a lesser known character from the Sesame Street franchise known as Telly. A fuzzy, purple creature with a large orange nose, and antennae poised upon his head, Telly was the pogo sticking, triangle playing, worrywart, proud owner of a pet hamster named Chuckie Sue. If imagination could heal hearts, Isabella's family and friends just turned on the proper road to recovery.

On February 10, Stacey launched the web page with the first update on Isabella's condition.

Isabella has a rare bacterial infection known as Gas Gangrene. The doctors are not sure how she developed this, but think it may have something to do with a weakened immune system. She is still very ill and will be on the ventilator a long time, and will need multiple skin grafting in the areas where they removed muscle and tissue. The doctors seem a little more optimistic about her recovery, although she still is very critical and things could change at any time. They don't want to give us false hope. Today was her best day so far. We ask that everyone pray for our little girl and add her to your church's prayer list that she might pull through this illness.

Silence crawled upon the pale walls within the isolated, clean room that now stood as both Isabella's sanctuary, and prison. The astringent odor of disinfectant clung to the air and became unsettling to the stomach. At times, it seemed that the sharp steel needles intruding Isabella could be tasted; bitter and metallically blood-like. Despite the softening effect of the diffuser panels, the overhead fluorescent tubes cast lines of sterile, pallor light. They hung in perfect sequence as if conformity would somehow diffuse the chaos of a failing body. For Keith, it stood more as a reminder of an order of things which were not of his making. Quietly, he sat alone next to his daughter's bed.

The stillness that permeated the hospital room was far removed from the tornado that raged in his mind. It cast disturbing images about like debris that impaled any pleasant memory which dared to present itself.

Isabella—Bella—Bell—The Bird—Sweet Baby Girl—Daddy's Little Daughter. He and Stacey had been elated when they found out they were having a girl. They had many conversations and Stacey had made her hopes well known. She longed for a girl; a female companion of her own making.

Sometimes a man can feel pressured to perform in a situation of which he possesses no control. Knowing how to have a child was easy, but Keith was uncertain how to avoid putting the proverbial "stem on the apple." To his good fortune, God took care of that detail.

Isabella arrived with anticipation tantamount to entering the Salle des Etats at the famous Louvre museum in Paris. While Keith was no art connoisseur, he knew a Mona Lisa when he saw one—the demur features that demanded attention. Isabella was he and Stacey's magnum opus; a masterpiece which would require delicate handling to ensure proper preservation.

On the night following her birth, Isabella was wheeled into Stacey's hospital room and left with her groggy, freshly incised mother. The next morning when Keith arrived he found Stacey frustrated to tears. She told Keith that the staff had brought Isabella in and left her. In her youth, Stacey gained no experience as a babysitter. In fact, her first exposure to an infant in her care was with Isabella. She held her newborn infant in her arms, petrified, without a clue of how to actually care for her. Keith immediately went to the nurses' station to inform them he would be staying the night with his wife and child.

The receptionist said, "I'll check with my supervisor and see if it's alright."

"You can check all you want," Keith responded. "But you're going to have to get security to remove me. Even if I have to sleep on the floor, I'm not going to leave my wife alone again."

Once at home, he and Stacey spent nights pacing the floor while Isabella howled from the pain of colic. They spent days monitoring the child's irregular bowel like storm chasers waiting for the class five twister. It was a case of the 'Terrible bullies,' a belly ache spawned by a lack of bowel production that could linger for days. Keith and Stacey found themselves huddling over the changing table, praying for the odor that would lead them to a celebrant poop; only to find an empty diaper leaving them to resume their watch. During one such vigilance, they were administering a doctor-recommended suppository when the poop shot across her changing table, startling them both. They celebrated with a chorus of "She pooh-poohs in her ditty," a song they had made up for just such an occasion. Just like simple things become large problems in the minds of parent, simple things can also become tiny triumphs.

With the thought of her stomach pain, another twister surfaced. The Williamsport ER, the very hospital where Stacey worked was now ground zero for a day of indescribable horror. The blackening flesh. The spewing blood. The terror in Isabella's eyes. The moments were etched in his mind like a tattoo. The ink of a parent's worst nightmare. His heart was breaking and re-breaking with every scene that he couldn't seem to escape. His mind was forcing him to endure another replay of that night, as it had since their arrival in Pittsburgh. Syndicated for repeat in full color with surround sound, it played out. He didn't want to see her face. He didn't want to see their faces. He didn't want to hear those sounds.

He pushed it from his mind, struggling to remain positive. This was a daunting task in a place where the death of a child can occur at any given minute. The little seven-year-old girl, who had originally arrived with a fever and pneumonia, had been given a new diagnoses, leukemia. The prognosis was bleak and family members were being called in from across the state to say goodbye. Here, lives didn't come to a poetic, Shakespearean ending, nor did the end come in graphic Hollywood style. It was a cold disconnection. The announcement of a time of death. Official. Nothing left but the paperwork. Keith knew hospitals had to manage this way, but with the light once again shown upon the function of the machine, the shadows of the business side of healing were revealed in all of its haunting irreverence.

He had been there. He too was a witness. At Geisinger Medical Center some fifteen years ago, when he had walked into the room to see his ex-wife Lori holding their little Samantha as the life faded from her skin. When Keith stepped in to say goodbye, a nurse wrapped Samantha in a blanket. Keith sat with the body of his daughter in his arms. He pulled the blanket up to her nose. It looked like she was only sleeping. But the lie couldn't hold back his tears. There in a room with a half-dozen healthcare professionals, he was alone with his Samantha. He kissed her forehead. The warmth of her body had already faded. He nestled her to his chest and held her tight one last time. She was just sleeping, right? He wept for what seemed like the first time in his life.

A few months after her passing and still in the midst of a deep depression, Keith had a dream in which he was walking hand-in-hand with Samantha. The sky was as blue as Samantha's eyes, the day was bright with afternoon sun and the sidewalk was unmarred by wear. They walked a while before he looked down at her and asked, "Are you okay?"

"Yes," she answered.

Keith awoke crying from longing and relief.

Samantha's name meant 'told by God'. Isabella's name was defined as 'consecrated to God.' Perhaps Samantha was the messenger that Isabella would be in God's service.

With great effort, he conjured up another positive image; Isabella with ice-pop-stained teeth laughing at him because, "Daddy's ice pop broke."

He rested his elbows on Isabella's bed and clasped his hands together near his lips. "Our father, who art in heaven," his voice came as a damp whisper. He had never been vigilant with any study of theology, but he had managed to learn the Lord's Prayer.

In the gospel of Saint Luke, Jesus taught his disciples to pray. He had also encouraged them to repeat their requests to God as often as needed. Jesus had presented a parable of a man waking up a friend in the middle of the night to ask for three loaves of bread. The friend denied the man's request until the man became persistent, asking for the bread time and again. Finally, the friend complied with as many loaves of bread as the man needed. 'Ask, and don't be afraid to keep asking.' His parable taught them to never swallow their faith after a single request.

As the words of his prayer passed his lips, his mind played an image on the blackened canvas of his tightly shut eyes. He was arriving home from work. Isabella had been waiting for his return and a simple ritual which brought her great joy. With a smile bright enough to ignite a heart, Isabella yelled "fan." He lifted her up and sat her on top of the refrigerator before turning the ceiling fan on with the countdown of three clicks. It was a tiny request that she would ask for again and again.

As the memory ignited a desire for the past, he prayed for Isabella's healing and for mercy to be bestowed upon the other tiny lives which hung in the balance. A quiet moment of desperation. A rare moment of hope.

A tickle on his cheek drew his hand. It came back damp. He swiped the tears away with his fingertips, unsure exactly when he had started to cry. He straightened himself and brushed aside some matted curls to expose Isabella's forehead. He pressed his lips to her skin. She was warm; another tiny triumph.

Entering the room, a nurse in her mid-twenties approached Keith at the bedside and looked over his shoulder at Isabella, who was resting peacefully in the arms of the life support and pain medications. "How's she doing?"

"She's doing well. She's tough. Tougher than her Daddy." He flashed a weak smile over his shoulder at Tiffany. Of Isabella's many nurses, Tiffany was one of their favorites. From early on, she had displayed a genuine compassion for the child.

"I'm going to inject her IV with a dose of additional pain medication. We will be changing her dressings in a few minutes."

"Okay." I'll step out a while." Keith turned back to Isabella and gave her hand a squeeze. "Daddy will be right back."

Keith pushed through the double doors and stood in the hallway, on the proper side of the yellow line. Grumpy the Door Troll, who had chastised him when he first arrived, was not on guard. The woman whom sat in her place was pleasant—to a degree. He rested his weight against one of the bland walls, wondering why they had not modernized the structures appearance. He was certain the lighting fixtures were on a dimmer switch which was being turned down a little each day for the sole purpose of enhancing his misery. The ceiling seemed to have dropped a little every time he looked at it. He was already feeling trapped in a place and situation that was slowly tapping his strength, but now these hallways were making him feel downright claustrophobic.

Stacey had gone to their hotel to shower in hopes of reviving somewhat. She would return within the hour. And although they shared in this suffering, he was happy to be alone for awhile. This was his time to spend inside himself; time to think.

A wheelchair approached carrying a girl of no more than ten-years-of-age. Her body was profoundly twisted and made her appear to be in a position of grave discomfort. A man who Keith assumed to be her father, guided the rolling chair with a woman walking at the young girl's side; probably her mother. Keith made certain to pull his gaze before it could be misconstrued as a stare. It was impolite to stare. Though he couldn't help but wonder of Isabella's condition, God willing she survives this ordeal. The family passed him, directing themselves toward the receptionist's window with purpose.

The receptionist greeted them not only warmly, but with genuine enthusiasm. The young girl in the chair had once been treated at this very ICU. They discussed the girl's success and her continued treatment away from the PICU. And though they hadn't put it in words, Keith could hear the gratitude in the voices of the parents. As they continued their conversation, he felt an agonizing change inside; a pain far deeper than the body. He could feel his soul writhing from within. All of the feelings he once hid when he saw people with serious afflictions, feelings of aversion and separation, were now blatantly shameful to him. In the eyes of that little girl, he saw his child and in the eyes of the parents who loved her for what she was, he witnessed the face of God. He wanted to hug the parents and kneel next to the child, hold her hand and ask for forgiveness. Yes, this place was filled with human suffering caused by human problems, but God sleeps here with all these children surrounding Him. He walks in the halls with grief stricken parents and flies with the flight crew of the life-flight helicopter to bring more sickness and suffering to a place that has it oozing from its foundation. Keith was comforted to know that He also sits with him at Isabella's bedside while he holds her hand, reads her a book or whispers, "Daddy's here, sweet stuff, Daddy's here."

He had called this hospital a God forsaken place, a comment for which he was now compelled to apologize.

It was his choice. He could choose to view the suffering or see this hospital for the miracles that walk in, and, as in the case of this little child, walk out of this place of healing. Suddenly, the little girl was not in a wheelchair after all. She was being carried by a man who left no footprints.

Keith could feel his spirit as it rose in his body and lifted the hairs on his skin. And though he knew his legs would have been wobbly without the support of the wall, he could feel a surge that nearly convinced him of the possibility for unaided flight. When an oppressive burdened of weight is lifted, you feel lighter. Fleet-footed. Boundaries of laws and physics need not apply here. And in that moment, the ceiling raised to its full height, and the dimmer switch was turned to its maximum intensity.

Nanny was born on August 2, 1934, and began her life as Joyce Arlene Kinney; daughter of Edward, who was the great-grandchild of Isaac McKinney, an immigrant who came from Ireland to the United States where he later dropped the "Mc" from his surname. Traversing dark fathoms, the ship upon which he traveled was undoubtedly guided by the stars; its passengers brazened by the hopes of finding a new life. Nanny's Irish blood had been slightly diluted with a mixture of a few other genetic heritages; but the Irish within her ruled by majority in her family tree, in her DNA and in her heart. And perhaps, it was also the root of this fair-skinned, grey-eyed, brown-haired lass's penchant for guilt-ridden augury, a condition, for which the Irish were well documented. She monitored her own actions, fearful they may become the spearhead of a self-fulfilling prophecy. Removing Isabella's toys from their normal place in the living room had put her Irish at odds with herself. She had abandoned the child. She had put Isabella's toys away; thus she was accepting that Isabella would not return; and therefore making her culpable if the child did not recover. Troubled by her own actions, Nanny pondered her decision, even discussed the matter with her daughter-in-law. And unlike the voyage of her great-great grandfather, Isaac, she had no stars as guidance—return the toys to the living room and agonize over Isabella's absence or leave the items out of sight and be ravaged by guilt.

Fortunately, her self-oppression comes with a hidden treasure for the tortured soul. A golden nugget buried beneath the bedrock of a rushing stream. Though not within the desperate prospector's sight, it glimmers with a light that can only be felt—the light of hope.

As Isabella is the sister of Samantha, hope is the sister of faith. And the greatest test of one's faith is to believe goodness, in its darkest hour, will triumph over evil. Clearly, if any evil had ever had a protagonist in their hour of desperation, gas gangrene, caused by what the doctors called Clostridium septicum, was responsible for the gravest challenge of Isabella's young life.

In times of travel, we are guided by a collection of stars that, aided by our imaginations, form recognizable pictures for us. Or perhaps, we are directed toward a single star in the night sky; a small yet brilliant source of guidance against the pitch of darkness; providing a light that is the only difference between the comfort of familiarity and the cold, unforgiving void of utter nothingness. For Nanny, that small brilliant source of guidance was Isabella.

With the western hills now ablaze by the reddish-orange embers of the falling sun, the night sky settled in for its task of ushering in a new beginning. The beacons of the celestial sea began to flicker on and the moon held dominion over the rise and fall of the tides. In a few days, the daughter of a simple Irish immigrant will embark on a journey to Pittsburgh, following the distant but ever illuminate polestar known as Isabella.

Nanny returned the final toy to its original position in the living room and prayed that her diligence would be rewarded.

February 13, 2005

Two hallways merge into a T on the second floor in the main tower of the Children's Hospital of Pittsburgh. A turn to the left placed Keith and Stacey in the middle of dueling retreats; a quaint gift shop and the hospital's chapel. The door to the right, akin to the others used in the hospital but stained to a smooth walnut finish, opened into a tiny chapel of eight pews. Like a child aspiring to follow in the footsteps of a parent, the chapel was modest in its homage. The room featured a high-volume ceiling, though not of the height and majesty of its vaulted, cathedral counterpart. Instead of buttresses, Rosetta trim and paintings, the ceiling boasted an unremarkable brass chandelier. Instead of stained-glass windows with brilliantly colored saints or an artist's rendition of his favorite biblical verse, the windows were covered with white interior shutters. The altar was little more than a boxy alcove painted glossy-white and appearing as aseptic as any operating room. An organ was nestled against the wall to the right and a closed communion table, with a spray of plastic flowers as its centerpiece, was pressed against the far wall. Instead of a crucifix, a wall hanging displayed a print of the ubiquitous Lindberg painting of a guardian angel watching over a young boy and girl as they crossed a rickety, slat bridge.

In modern times, a chapel within a hospital is as common as was an infirmary within an ancient monastery. However, long before there were doctors, there were witch doctors; before the establishment of medicine there were medicine men. Before mankind ever considered treating the body for an illness, they attempted to heal flesh by treating the soul. Mankind's first venture into surgery was based upon spiritual healing. In man's most primitive state, they attempted to release the demons that caused ailments by such brutal methods as trephination. Using an instrument called a trephine; the medicine man would remove a circular piece of bone from your skull. Whether your infirmity was diabetes or leprosy, nothing could release an evil spirit like a nice hole drilled in your head. But through knowledge, civilization had arisen.

The ancient Romans and Greeks utilized their temples to treat the sick and the wounded. In India, the first hospitals were established during the time of Buddha. But it was the life of Jesus of Nazareth that spawned the largest expansion of hospitals in ancient history. In the first thousand years following his death, the Christian Church built hundreds of hospitals in scores of nations. All constructed from the inspiration of a carpenter. Healing the body became an integral part of an organization created to heal the spirit. An awe-inspiring physical and psychological symbiotic partnership was born—a single sanctuary built on the steadfast foundation of a rock called medicine and a rock named Peter.

Medical facilities were a rare arena that had not had God completely expunged from its infrastructure. Somehow, the society that managed to turn a pagan-despising bishop and later saint named Nicholas into an icon of paganism called Santa Claus, had been unsuccessful when it came to hospitals. During his lifetime, Saint Nicholas was known to have such intolerance for pagans that he was drawn to fists. Upon his examination of his current iconic personae, we may get a pass on giving his lean figure an extra two-hundred pounds to carry. He may forgive us for stripping away his bishop's cloak for a jolly, red snowsuit. He may even excuse the holly switch that we often place in his cap. But certainly when he noticed that we failed to emblazon a holy cross on his new vestments, he, Santa Claus, would box our ears; ceasing only when he discovered that we managed to leave a chapel in our hospitals.

In almost any given hospital, there is a hallway which leads the despaired of heart and the desperate of mind to the hospital's chapel in hopes of salvation. Here, prayer transforms the undertow of murky uncertainty, the gashing rocks of the unexpected and the suction of sheer vulnerability into a silvery sphere of ascendancy. A buoyant sphere which raises the genuflecting soul out of the swirling waters, up into the fresh air of function; no longer helpless as the doctors work to save their loved ones, as prayer provides participation in the healing process. The parents, who cared for and protected their young, who now found themselves relying on others to nurture their offspring, have but a single remedy at their disposal during times of severe medical crisis; prayer—and they clutch it like an infant to its mother's finger.

It was Sunday morning as Keith and Stacey entered at the rear of the chapel where a small table pressed against a wall held a simple three-ring binder filled with lined notebook paper; though the hand-written entries were anything but simple. The binder was a prayer book for the parents and loved ones of children under the care of the hospital. With Keith hovering above her shoulder in concentrated interest, Stacey flipped through the pages and read several of the pleas for God's mercy. The ink deducibly quivering with each hand-stroke, leaving Keith and Stacey with the sense that the pages also bore the salty stains of the woeful as they labored to maintain their focus through aqueous eyes while recording, for God and man, the life and death struggle of a child.

Like frightened and confused castaways stranded on their own private island of parental pang, Keith and Stacey had waited with their eyes heavenward in hopes of being rescued. They cast up prayers of being whisked away from the entrapment of their isolated ill wind, so they may return to the normalcy of mundane living where troubles meant how to get the child to preschool without being late for work. And all this time, despite the precious young children who clung to life in the beds encircling the pediatric intensive care unit, Keith and Stacey had felt alone in their suffering. But as the prayer book could attest, this hospital was an ocean teaming with enclaves occupied by tormented loved ones.

Stacey wrote a prayer for Isabella before sitting with Keith in one of the pews. A smaller, framed version of Lindberg's cookie-cutter, angel painting watched over them. They hoped the blessed angel portrayed in the print continued to watch over their daughter.

The picturesque town of Chester, New Jersey had somehow managed to fortify its roots in a cultural forest of clear-cutting. Chester was a Norman Rockwell postcard with many of the original buildings in the town-square on Main Street still servicing the public. It was quiet, undisturbed by the six lanes of traffic which thundered along the final stretch of Interstate 80 not far to the north. Through the industrial revolution, and the technological advances of the modern age, the denizens of the town once known as Black River, were still the same God-fearing, English Puritans who had settle in the rolling hills of Morris County in early eighteenth century America. And within thirty years, they had formed a church that would last through the centuries.

The parking lot blacktop was clear, but the surrounding earth was crusted with a mantle of snow crystals compressed from several afternoons of melting followed by evenings of freezing. The snow had gone from granular to a single, well-polished tile that reflected nearly all of the sunlight back into the ozone. The sky radiated azure blue through the thin air with only an occasional wisp of cloud strewn among the sky like angel hair brushed upon a dampened blue canvas.

To some members of the congregation, the needles of crisp cold that all but penetrated the skin served as an exhilarating awakening on a lazy Sunday morning. They stood outside of their cars on the southern side of the First Congregational Church and caught wind of the life events that carried their fellow parishioners across the previous week. Some conversations were privy to only a few, while others were intended for mass consumption. A father spoke of a less than enthusiastic teen wielding a snow shovel who was convinced that sidewalks were not only unnecessary, but posed a health crisis to today's youth. The teen parried with a public request for a snow blower. The father counter parried with a brief lecture regarding the benefits of manual labor and a quip about idle hands doing the devil's work. The teen offered a riposte in an impromptu quote, stating that an idle wallet does the tightwad's work. The gathering chuckled, sending plumes of vapor skyward as the father casts a wily smile and a "touché" wink at the teen for a duel well fought.

Those not of the mind to tolerate the elements and those whom decided their time was best spent in quiet communion with the Lord had already retreated inside the boxy wooden structure peaked with a double-pitched roof and an old, modest belfry.

The signature of a strong church laid in the symbiotic nature of the relationship between the organization (considering that the organization can count God among its proprietaries) and the members of the congregation. Each thrives on the other's success until one cannot exist without the other. The First Congregational Church of Chester was autonomous and staunch in the practice of

God's word. And with well over 200 years of practical worship under their belt, they had become a sentient entity in which the mind served the cells, and the cells served the mind.

The face the assembly displayed was as recognizable as that of any of its members. It was a perfect blend of the faces of all who had worshiped under the call of the order and of the structure wherein the members consecrated their faith. The church had become the chapel. Humble in its appearance, natural in its origins and strong and unwavering in its foundation.

With painted slat-siding as pure white as the crystalline granules it nestled in, the chapel seemed to monitor the parking-lot crowd with two sets of large eyes; windows of nine narrow, glass panes, each nearly five feet in height, stacked in three layers of three. Each fourteen-foot window was accented with an eyebrow made from an arch of black, half-round molding which wrapped around the frame before tapering into three concentrically diminishing half-rings that formed the point of an earth bound arrow.

On the east side of the building, a small courtyard bore two black, iron lampposts and a sign proudly announcing the First Congregational Church and an underlying earmark for a Christian child daycare center which operated out of the old church meeting hall. The courtyard trees hibernated with their leafless branches spread out awaiting the warm kiss of spring. A wooden bench sat poised beneath the youngest maple for any member in need of a place for quiet reflection. To the north, the monuments of the Chester cemetery peeked through the evergreens that lined the edge of the courtyard.

A concrete walkway dissected the lawn before leading up to a portico with four white columns, each nearly the height of four men. Both the number of pillars and the number of windows on the south side of the chapel signified the four Evangelists (Mark, Matthew, Luke and John). Double oak doors painted hunter green lead into the narthex. A sign above the doors read: "1st Congregational Church. 1st House of Worship was erected in 1747. 2nd in 1803. And the 3rd in 1856. Enter into his gates with thanksgiving, and into his courts with praise. Offer the sacrifices of righteousness, and put your trust in the Lord."

Inside the narthex, the end of a thick rope coiled on the floor to the left of the entrance before threading its way skyward into the second level and at last into the bell tower. The church bell was rung when the service began, following the Lord's Prayer and upon special occasions such as the presentation of a bride and her groom as they departed the sanctuary.

Beyond the threshold of the narthex, a spacious nave of white and gold greeted the congregation with four rows of pews. The windows of the Evangelists watched the proceedings from both the north and south walls, draped in white sheers and accented with golden valances. Between each window a column was painted on the wall, the first signs of the Trompe-L'oeil which lent depth to the boxy room. A French term meaning "trick the eye," the three-dimensional art technique deceived the viewer into believing the column protruded from the wall.

A lectern stood alone in the north corner of the sanctuary. In the southern corner, railing hedged an 1873 tracker organ which featured ivory pipes trimmed in gold and teal. The rail continued forward along a low stage that held a small organ, congas and various musical accessories. A wooden pulpit at the center of the sanctuary was emblazoned with a gold cross; another

Trompe-L'oeil painting of an apse provided an appealing extension in the backdrop.

When the parishioners had finished filing in, a tall, lean man stepped to the pulpit to deliver his sermon. The hair of his mustache and temples showed his wisdom, his eyes displayed his compassion. Pastor Hoffman welcomed those in attendance, led them in hymns, spoke to them of Jesus and preached to them the word. He led them in the Lord's Prayer, and then paused while the bell tolled high in the tower to ring the message beyond the chapel. Pastor Hoffman then asked for requests, an open invitation to raise the voices of the congregation for anyone in need.

Bob, the deacon of the church, made a request in proxy for another member of the congregation. He was a personal friend of Don Brown, another member of the congregation who was unable to attend the service due to an illness in his wife's family. Bob and Don had become jogging partners shortly after cancer claimed Bob's brother. Don's niece, a two-year-old Pennsylvania girl, was fighting for her life with a grave illness. She was already on the Prayer List for private prayers conducted by volunteers, but Bob requested a public prayer be made for a girl named Isabella.

The ancestors of English Puritans, surrounded by the spirits of those whose convictions they upheld, lifted the little girl up by name; asked God's grace in healing the child; asked for discernment for the medical staff that attended her and asked for the heavenly angels to surround and protect her. And with the resonance of the church bell still echoing across the Black River valley, they prayed for a child they have never met, yet knew her by her name; knew her as a child of God; and knew her by her plight.

February 15, 2005

The news regarding the existence of Isabella's 'Care Page' webbed out across the network of family and friends. During these days, information seemed to be in short supply, oftentimes enigmatic and never sufficiently swift. Not to anyone's fault. The distance, the strange nature of her affliction and the fact that the parents remained primarily at her side, were all roadblocks in the information highway. Now, updates would be as quick as the click of a mouse. The hospital had provided the internet server and *Sesame Street* had provided the inspiration for the username; 'Telly.' Although, it was a full five days before the vehicle known as 'Telly' cleared the traffic jam to become part of the common vocabulary for those in the need to know. Internet travelers began pouring into the site.

The first entry came from one of Keith's co-workers on February 15th.

> *I don't think my first message went through. Have been and will continue to pray for you and your family. I have put you on the prayer list of two churches. There IS power in numbers! Hang in there! Never give up hope! I was glad*

to read the message today! I know you have a tough road ahead! You have a lot of people pulling for you.

Let us know what we can do to help! I will check this page daily for updates. May God hold you all in the palm of his hand.

Vicki S.

Every highway has a point of origin. Mile-marker one was at the Children's Hospital of Pittsburgh, where it was the pieces of information that did not pertain to Isabella's condition that were not shared on 'Telly" that added weight to Keith and Stacey's sagging spirits. These were stories of the suffering of others around them.

The PICU waiting area had become a community of anxious neighbors. There were moments of comforting triumph when a family moved out of intensive care with a child well enough to go home or, at least, out of imminent danger to allow transfer to a less restricted area of the facility. And there were moments of loss when a family's stay ended with the words, "I'm sorry."

The middle-aged man whose wife had left him alone to care for their twelve-year-old, wheelchair bound boy was told by physicians that there was no more they could do for the child. The child was sent home to die. Though weary in spirit, the man who had dedicated his life to the boy would be there with him until his last breath. They departed the PICU and the "Waiting Room Community" never heard from them again.

The Community was present to witness a marriage proposal, something you don't often see in such a morose setting. While the seven-year-old girl with leukemia clung to life, her mother's boyfriend had asked her mommy to marry him. No one was sure how the little girl would feel about having him as her new daddy, and that would be a secret she would keep. The little girl whose pneumonia turned out to be cancer quietly passed away a few days later. A rumor began to circulate that the mom had been interviewed by a local news channel. The little girl never got the chance to appear on that local TV channel. Instead, she was appearing in heaven.

As for Keith and Stacey's family, Isabella was the most qualified TV personality. The day after she was born, a local news station came to the Williamsport Hospital nursery to gather film footage for a story regarding unwanted babies. Isabella was taped as the counter-story. Basked in the bright lights of the camera, Isabella was shown nestled comfortably in her crystal bassinet as the reporter proclaimed, "This little girl is lucky enough to have a loving family to go home to." Of course, a copy of the newscast was captured on tape and placed in Isabella's baby cinema library.

Beyond the untold tales of the other parents, there was another storyline which was tactfully withheld from Keith and Stacey's posted updates on the 'Telly' web page. A story which would leave Keith grappling in the darkness of his blind faith in the medical field and have Stacey directing light-shedding fireballs at the truth behind their daughter's care; and once again, exposing Keith's angels of healthcare as evil head-eating aliens.

Her eyes were mostly shut. Oily discharge thickened her lashes. Her body was still, but for the rhythmic heave of her chest. Are you in there, Isabella? Are you quiet to the pain?

Keith busied himself by looking at the many 'get well' cards that had been taped to the far wall of the isolation room. Some mentioned prayers. Some had words designed to somehow evoke Isabella's inner strength. All declared their love for the child.

Nanny was set to arrive later that afternoon. Keith was thankful. Besides Stacey, if there was anyone in this world Isabella would respond to, it would be her Nanny. As for he himself, you're never too old to feel comforted by the presence of your mom.

Stacey sat in a leather bound chair next to Isabella's bed. Comfort eluded her. Unlike Keith, she viewed Isabella's care with skepticism even though she was a nurse, and probably because she was a nurse. The pressure of having the knowledge, but being nearly powerless to utilize it, gave her a false sense of responsibility. If anything happened to Isabella, she would be crushed beneath her own guilt. Knowledge can trump faith, but with knowledge comes the same accountability those of the faith place on God.

In her family, success was not an option, it was a necessity. Her father, Jim, instilled that into her. You had to demand what you wanted. In the world of finance, it served her well. In the world of family dynamics, it often left her floundering and off-balance. When you have something to sell, it gives you bargaining power. When you have something to give, there can be no conditions.

As her primary teacher, Jim became her pillar of strength. The one she would count on in any crisis situation. Several years ago, that pillar was diagnosed with multiple myeloma. The man who had always been there to pull her to her feet when she fell had cancer.

Jim's brother, a doctor in New Jersey, had always served as his primary physician. However, the cancer warranted the need for a local doctor. There was great uncertainty in their selection. And by proxy of her profession, Stacey was an integral voice in the family's decision. The microscope of skepticism was brought out and anyone studying Jim's cells would likewise be studied. Now, with Isabella, she was battling her colleagues on two fronts. She had never lost a family member and she would do anything in her power to make certain it didn't happen.

To this point, Stacey's only real loss was two dogs. Kiko and Shanghi were Stacey's children in her pre-Isabella world. But they were Shar-Pei, a breed notorious for having health problems which ultimately forced Stacey to have them euthanized. The loss of the dogs was painful. Jim's illness rocked her to her core. Isabella's condition brought unimaginable pain. Hopelessness. Depression.

And a pressure that someone with utter faith in the healthcare system may never understand.

While Stacey was a product of modality, Keith was a product of himself. He was raised in a family of love and acceptance. Not simply the acceptance of human folly, but acceptance of their status in life. His father avoided failure by means of non-participation. His mother accepted failure by means of love.

From the time of his youth, Keith's father, Ralph, approached life with both a gregarious and pugnacious zeal. As a teen, he joined the high school football squad only to quit before his first game after refusing to obey his coach. Despite being a musician, a gifted singer and an avid fan of country music, Ralph never pursued music as a career or a means of performing for live audiences. He was given an opportunity to front a band, but during a performance at the local Veteran's Club when he was faced with the pressure of having his talents judged; Ralph played the entire set with his back turned toward the audience. His friends shrugged it off as one of his typical social snubs, but it was his fear of failure and an overwhelming insecurity which caused him to shutdown. He would settle for performances at parties, with other family musicians, to showcase talent.

Ralph had always elected good times over good fortune. He never owned a home, spent thirty-five years as a worker in a boiler making factory where he had an average attendance record and spent most of his spare time and money on weekend binges. Joyce stayed faithfully at his side, raising their children, cleaning their house and tolerating his antics. As a result, many of their eight children had fallen victim to complacency even after striking out on their own. Live modestly and remember that success comes at a price.

From an early age, it was obvious that Keith was different. His desire for grandeur displayed itself in sometimes comedic and more oftentimes brooding manifestations. He possessed his father's tenacity but portrayed it by means other than violence. And like his father, Keith had a gift for music, though later he pursued his dream throughout his adult life while his father had not. Keith abhorred the path of dependency that his father accepted. Following a brief stint as a tenant during his first marriage, he vowed to always be a homeowner. He aspired to advance his career in music and his eventual career at the fast food restaurant where he had worked since leaving high school. As well as expecting more from life, Keith expected more from himself. He was critical of himself to a fault and oftentimes his obsession with his own foibles prevented him from embracing his strengths. His self-cynic continued to clash with his self-esteem, but a slow change was on the horizon; brought about from an unlikely source.

Stacey had always been driven to succeed where so many others were left drowning in debt by their own apathy. She was relentless in directing Keith toward financial efficiency and independence. At times, Stacey became his worst critic, accepting the onus he had once entrusted to himself. And by breaking him, Stacey was inadvertently repairing him.

The opening of the door drew Keith's attention as a pair of technicians entered the room. Both women were dressed in blue uniforms, one had what looked like a large, metal flip-chart tucked under her arm.

"Hi," one of them addressed Stacey. "We just have to do a chest x-ray."

The flipchart turned out to be an x-ray board which the technician placed at the foot of Isabella's bed, while her counterpart swung the large, white arm of the in-room x-ray machine in place. The pair moved with mechanized fluidity as though they had performed dozens of these, as a team each day. Like the refined parts of a grasshopper oil pump, each set about separate tasks, yet moved as one.

"Are you aware of her condition?" Stacey asked.

The pump stopped. Basked in the glow of an all revealing fireball, the technicians shared a confounded gaze.

"Do you know why she's here?" Stacey asked again.

They suddenly seemed shocked that their obvious talents as a grasshopper rig were being called into question.

"Until you discuss her condition with Amanda," Stacey referred to the nurse on duty, "you are not proceeding with the x-ray."

No words were exchanged by the pair, just a disgusted sigh. They marched out of the room and into the anteroom, looking less like a well-oiled machine and more like thickset, bumbling bookends.

"They were going to shove a hard board under her back without any idea about her wounds." Stacey continued. "Look how unsympathetic they were."

Keith could understand Stacey's point, but it didn't make the situation any less discomforting. He felt defensive for Isabella, yet couldn't help but feel a sense of pity for the technicians.

The pair reentered the room following a brief discussion with Nurse Amanda. Stacey had won her battle but was still in the mood to explain her actions. She proceeded like a parent who had just scolded her children and now wanted to push the point home by explaining their need for discipline. You brought this upon yourselves.

"My daughter is suffering from necrotizing fascitis and has been undergoing debridement on her back and right leg. In fact, they're still cutting away dead tissue. It's raw. There's no skin just bare flesh and bone. It's blistered and still oozing. And I don't want her to experience any more pain than she has to."

"We understand," one of the bookends replied. "We are going to have to ask you to step outside for a moment due to the exposure."

"No. I'm staying in here."

"That's fine. We have a lead apron for you."

"I'll step outside," Keith said, as he headed for the door, leaving Stacey alone with the stout technicians. He admitted to himself the pair had acted a little calloused, although they were just doing their jobs. Stacey's ability to expose the staff's mishandling of Isabella was valiant on her part, but to him—it was unnerving. At least these two didn't turn out to be carnivorous aliens.

Stacey kept a watchful eye on Tweedle Dee and Tweedle Dum as they took their x-ray. She imagined if she found their lockers, she would discover a matching set of multicolored, brimmed, propeller beanie caps.

She understood that x-rays were a routine, noninvasive test, but the patients comfort should always be the primary concern. Of course the blame wasn't all theirs, there was a definite breakdown in communications between the staff, and Stacey was getting sick of it.

The face was recognizable, but not altogether familiar. The lips, dried and puffy, were curled out as though she were playfully imitating a fish. Her face was abnormally swollen due to, as Keith explained, the steroids being administered to help speed the healing process—things Nanny didn't fully understand, though accepted. The tubes running to and from her body and the environment created by the surrounding monitors completed the cruel disguise. At least the wavy, auburn hair was undeniably Isabella's. This was Nanny's granddaughter, just not as she remembered. The child's suffering was painfully apparent in her current condition. The capitulatory plunge of Nanny's heart into her stomach nauseated her. Shimmering droplets clung to her eyelashes, collecting until a blink sent an accumulative drop coursing down her cheek.

Keith leaned down, kissed Isabella's forehead and whispered something in the child's ear. His eyes lifted to the digital displays on the monitors as he surveyed the numbers for any signs of elevated heart rate or blood pressure; the harbingers of pain. He explained how elevations in Isabella's vital signs meant she was experiencing pain; once again Nanny accepted this without fully understanding. The only word that registered in her mind was 'pain.' Her granddaughter was feeling the pang of these horrible injuries. Keith had once repeated the surgeon's descriptions to her; the flesh had been cut away until the surgical knives exposed bone—femur, spine and ribs. The uncontrolled bacteria had opened the gateway to Isabella's own personal passage through Satan's torture chambers. Calling Isabella's suffering "pain" was tantamount to referring to the mutilated souls of Dante's purgatory as "inconvenienced spirits." Pain was a boo-boo. Pain was something children got from falling in the playground. Pain could be kissed away. These were deep wounds! Isabella was in sisterhood with Hypatia, the curator of the Library of Alexandria who in the dawn of the Dark Ages had her entire body flayed of flesh by a mob of her detractors, using only abalone shells. Though Hypatia was fully awake during her ordeal; sedated or not, Isabella's wounds must hurt like hell.

The weight of the child's distress pressed against Nanny's lungs until she labored for every breath. *Why Isabella? Why this child?* The shaking of her hands returned along with the numbness of her lip; the anxiety she had so often battled was on the attack. Butterflies joined her heart as they fluttered to its rhythm in the confines of her stomach. She couldn't remain in this place for much longer.

"Keith," she said, "I'm going to go back out so Stacey can come in."

"Okay, Ma," he replied, "I'll be out in a second."

She hurried into the anteroom and removed her mask and gown. Exiting through the far door, she made her way around the gauntlet of critical care children with her eyes more often averted to the floor. A rare glance at some of the other patients who displayed the same swelling as Isabella brought her some solace. At least it was the medication causing the bloating and not Isabella's illness. The door leading to the PICU opened to the hallway where Stacey was waiting to greet her.

"Oh, God," Nanny managed. "She looks so sick, Stacey."

Stacey's eyes weakened, "I'm afraid this is too much for her. I've seen people hold on because they were afraid to leave their loved ones. I think we should tell her it's okay if she has to go; if she can't stand the pain anymore."

Nanny crossed the hallway and leaned against the opposite wall, her face melting away in grief. "I'll try," she managed through sobs more audible than her words.

Stacey lowered her head, "I don't want Isabella to suffer," she uttered before pushing her way through the PICU door.

Alone, Nanny struggled to regain her composure. The occasional passerby seemed perfectly accustomed to seeing a person immersed in complete emotional distress—natural behavioral boundaries as established by locale.

All wounds don't heal with time; they scab over and await the next disaster to pick them back open. Memories of disjointed time spent in a hospital while Samantha laid dying oozed to the surface and infected her optimistic view of Isabella's chance of survival. She hadn't the strength to tell Isabella to surrender. But what was right in a place where children, filled with edema inducing medications, have to look to machines and monitors for nurturing?

The opening of the PICU door drew Nanny's attention as Keith stepped out into the hallway.

"Are you alright, Mama?" Keith asked as he approached her.

Nanny's body now shook uncontrollably, "Stacey said I should tell Isabella that it's okay to let go."

"She told me to do that, too."

"I saw you lean down and whisper something in Isabella's ear back there. What did you say to her?"

Keith spoke softly but his face displayed stone resolution. "I said, 'Isabella, you fight to live for Mommy and Daddy." He gazed off into space with a mitigated shake of his head. "I understand what Stacey is saying, but I'm not leaving here without her. And I'll repeat that a thousand times until it becomes real. We have to remain positive. I know it's tough to see her go through this, but she's come this far for a reason. They said she wouldn't make it past the first 24 hours; she did. They said it would be a miracle if she didn't develop organ failure; well, that miracle happened. It's been a week now, and she's still fighting. She's going to make it."

He placed a gentle hand on Nanny's shoulder. "We have to try to stay positive for Isabella's sake. She needs to know that she's not alone.

"I may have to go back to work next week. I hate to do that to Stacey—I mean, leaving her alone to deal with everything, but we don't know how long Isabella will be hospitalized. I don't know what else to do. By the end of this week I will have used up all my vacation days. Actually, a little sooner than that. Two of the people I work with donated a couple of their own vacation days to me."

"Really?" Nanny commented. "That was nice of them."

"We've gotten support from a lot of people."

"I know. I can't believe Stacey's brother-in-law . . ."

"Don." Keith added.

"God, I'm not good with names. Don. I can't believe he paid for my hotel room."

"Don's been great. He's helping me stay positive. I just wish Stacey could be more optimistic. I think she knows enough about medicine to keep her scared. It keeps her looking at the situation from a pragmatic point of view. She watches the monitors; the numbers. She insists on knowing all the medications and the doses they give Isabella. She's fighting with the doctors almost every day about Isabella's pain. Maybe she knows more about the situation than what she's telling me. Or maybe she's trying to prepare herself for the worst.

"I know the statistics are not on Isabella's side. I think they said it was a seventy percent fatality rate. But every day she keeps surprising the doctors, and each one of these days are going to add up to weeks. As long as she keeps fighting, I'm going to hold out hope."

February 16, 2005

Optimism is not always a straight and level road. There are peaks and valleys. Sometimes, the ascension of a mountain grants us a view of what is to come, and we are comforted by the sight. During the stretches of valley, our faith becomes a navigation system that keeps our optimism on track. The following entries appeared on the "Telly" Care Page.

February 16, 2005 at 12:08 PM EST

Isabella is still hanging in there. She is in surgery now and they are taking small bits of tissue etc. to control the bacteria. We are all praying that that will stop soon. They are going to start feeding her via tube today or tomorrow, and she is still on the ventilator. There are pictures of her in the photo gallery for anyone to view.

February 16, 2005 at 04:13 PM EST

We finally got some good news; they did not have to take anymore of her muscle in her leg, right now, and they are hopeful to try to do some covering of her leg and back on Friday. We still don't know about her function, but as of right now, they are optimistic.

February 20, 2005

Discomforting quiet. The distance between them was already growing without a single mile traveled. As it were, only the width of a bed separated them now. They stood like pillars on either side of Isabella as though the bed was a shrine to some banished Goddess of Youth and they alone were charged with maintaining her memory until the blessed day of her return.

Looming was an afternoon surgery that could determine Isabella's future. Afterwards, Keith's brother would arrive to drive Keith back to Williamsport, leaving Stacey alone in a hospital nearly four hours away from the only other person who truly knew how she felt. Her discontent showed in her taciturn demeanor, however, Keith held onto the slim hope that she would eventually approve of his decision.

To mollify the situation, Stacey's sister, Tanga, was also scheduled to arrive in Pittsburgh, this time to spend the week during Keith's absence; though he knew it was only a finger in the dike.

Keith looked up from stroking Isabella's palm. "Do you understand why I have to go back to work?"

Stacey turned her attention to the monitors to maneuver away from the discussion. "Her pressure is up," Stacey commented. "I asked that nurse over an hour ago for more pain medication, now where is she?"

Keith could see the creamy skin of her youth folding away within the lines of stress. The look of despair on her face had aged her appearance almost overnight. It had been two weeks of late-night vigils, torturous angst as they awaited word of the latest surgery and restless sleep on whatever piece of furniture was available at the time.

"Stacey . . ." Keith began.

"What do you want me to say, Keith? That it's alright? Well, it doesn't feel alright."

"Fine. I'll quit my job, but if I don't get back to work soon I'm going to lose my job anyway. I can't expect Jim to hold it forever. He has a business to run. I'll quit my job and take care of Isabella, but when the money stops coming in, you'll understand what a mistake it was."

The invading bacteria had entered through Isabella's skin then advanced to oxygen deprived areas deep in her tissue. To halt its progress, all infected tissue had to be removed, which exposed sensitive vessels and nerves. Of primary concern was the sciatic nerve, which controls the function of the legs. On previous surgeries, they had to be mindful not to touch or mishandle an instrument that could result in striking the exposed sciatic nerve for fear of causing paralysis.

Isabella was now undergoing surgery to move two of her existing muscles in her right leg to cover the nerve and to compensate for the muscle that had been previously removed. Stacey was hopeful that when the plastic surgeons were finished, she would have answers to the questions foremost in her mind.

Dr. Shao Sejong, a plastic surgeon of Asian descent with stoic disposition, had offered vague answers with little hope during previous meetings. When Stacey had asked, "Will she ever walk again?" He couldn't say for sure. When asked, "Will she be able to sit upright without assistance?" He didn't know. "Will she ever be able to do anything more than simply lay in her crib?" and "We don't know if any brain damage has occurred, so is it possible that she will be a vegetable?" He couldn't tell, but kept repeating something about 'in the acute care setting.' Frustrated, Stacey had asked what the term meant.

He had been referring to intensive care, saying that there were limitations to the plastic surgeries they would perform. They were concerned with the survival of the patient, thus covering the wounds from infection and to allow for healing were their only concerns. Functionality would be addressed at a later date.

"Are you saying that you're going to cover her back with skin, and then later on down the road, the skin will be removed, the area where her muscle is missing will be built back up and then skin would be replaced to cover again?" Stacey had asked.

"Basically, something like that."

And so it went, Stacey created the bleak images and Dr. Sejong hung them in the Hall of Possibilities. Stacey was discouraged that the expert refused to provide any insight, and Dr. Sejong may have been wondering why this mother assumed that fortune telling had been a part of his doctorate curriculum.

But that was before. Before this afternoon's surgery. Before Dr. Sejong was joined by Dr. Fagnano at their current conference. Dr. Fagnano may or may not have had the surgical prowess that Dr. Sejong possessed, but he definitely exceeded his colleague in optimism and perspicacity. Dr. Fagnano was young, perhaps in his early thirties. He was a man of diminutive stature, fair complexion, a lean runner's physique and wheat-blonde hair. His smile was pleasant and his presence placid. He rested his weight on the arm of an overstuffed chair. Dr. Sejong stood by the door with his arms folded over his chest. Keith, Stacey and Keith's brother, who had arrived less than an hour earlier, were scattered throughout the waiting room.

Dr Fagnano explained that the muscle in the inner thigh had been previously removed down to a thin layer covering her femur or thigh bone. To compensate for the debrided muscle, they had performed both a Sartorius turnover flap and a rectus femoris rotation flap in which the muscles of the front portion of the thigh are moved over to the inner-front part of the thigh to cover the exposed area and to assist in future leg function. Dr Fagnano assured them the procedure was routine and offered good results leaving little or no deformity or loss of function. The surgery was successful. Now they had to monitor the condition of the muscle. It would need to turn pink and beefy.

The other issue was her back. The loss of her latissimus dorsi (the first layer of muscle on the lower back) on the right side could easily be compensated for by the patient during trunk movements. This muscle was commonly donated for breast reconstruction.

Though happy with the sudden flow of information, Stacey's interest peaked when Keith asked the primary question that had haunted her since the debriding had begun.

"Will I ever be able to walk with my daughter around the block again?" Keith questioned.

Dr Fagnano responded with a resounding, "Yes." And then he added, "I don't see why not. She'll never be a track star or throw the javelin in the Olympics, but other than those things, she should be just like any other child her age with a fully functioning leg. She will have some adduction because of the loss of her inner thigh muscle, but the leg will be functional."

Stacey could see the tears well in Keith's eyes and the lump in her own throat was born of relief instead of sorrow. Now they needed Isabella to heal before other complications set in.

After a quick meal in the hospital cafeteria, the trio made their way through the labyrinth of hallways to the elevator. They ascended to the 8th floor. Keith took his brother back to visit Isabella while Stacey remained in one of the small waiting rooms. When the brothers returned, she escorted them back to the elevator. It wasn't a last mile walk, but their impending separation was evident in their posture.

In the ICU atmosphere, thick with the chill of mortality, Keith could feel the presence of a different storm; the brooding clouds of discontent and the slashing rain of detachment; like a torrent in the midst of a meteorological depression.

He turned to face the angry sky; Stacey's smoldering eyes held little promise of clearing.

"Everything is going to be fine. I promise. I'll be back on Thursday night."

"You just need to leave," Stacey said. "I can't tell you it's alright. Just go."

He hesitated, sliding his palms down the length of his face, and then quietly slipped into the elevator to descend to the lower levels and the parking lot.

A blanket of falling snow and an abandoned stretch of road construction just outside of Pittsburgh slowed their progress home. The drive with his brother was a mix of cell phone calls (including a brief call to a tight-lipped, annoyed Stacey) and conversation ranging from the details of Isabella's condition and treatment to the positive influence Don was having on Keith's psyche. Immersed in the negative atmosphere of a children's ICU, Keith had found himself clinging to Don's "life is what you make of it" soliloquies. It was like having his own motivational speaker at his disposal and Keith was drawn to the positive charge like an electron. However, neither his departure from Pittsburgh nor the subsequent arrival in Williamsport lent Keith a positive emotion. This was not a homecoming. This was a return to the scene of a catastrophic event. The reality of the way things are and the memory of the way things were collide here. In Williamsport there would be neighbors, friendly faces with good intentions. He dreaded the pitiful looks he was sure to receive from people he knew, or the occasional bold questioning. He was all too familiar with the guise of commiseration and the bumbling need to say the right thing. Silently, he wished to avoid as many people as possible.

Some cities inspire. New York. Hollywood. L.A.

Williamsport aspired. It aspired to be a big city. It aspired to be a small town. It aspired to look toward the future, and to maintain the past. Some believed the only thing Williamsport inspired was stagnancy. Except for a few glorious weeks in the waning days of every summer, where the youth from around the world put on a show of determination and sportsmanship that would become the calling card for the city's denizens. "I live in central Pennsylvania. I live in Williamsport, home of the Little League World Series."

But on a cold winter's night, far from the gnat-tickled, lush-green fields of Howard J. Lamade Stadium where the games were played, the now frozen soil of this city only served as a reminder of Keith's last experience in the city—a stinging departure in pursuit of an ill child. As he pulled the car into his driveway, Keith climbed out, wrestled his suitcase from the trunk compartment and thanked his brother for the ride. Approaching his house, the interior was eerily dark; far from Stacey's propensity to drive back the night with candescence. These homeowners had moved on. They left behind their furnishings, personal belongings and ornamentation in exchange for a harsh lesson in the priorities of life. Whoever these people were, they were not coming back; they would never be the same.

The despair of both of his daughters now weighed heavily upon his shoulders, creasing his brow and dampening his eyes. The sisters seemed to be reaching across time desperately trying to convey a message to their father. Keith gazed toward the heavens. The sky was crisp with the remnants of a few storm clouds drifting in the thin, cold air. A half moon hung above the rolling blue mountains toward the southwest.

Samantha loved the moon and would often peer out of a window for hours with her Daddy as she pointed out at the night sky and said "moon." One night she was having trouble sleeping and Keith opened the curtains so she could see the moon from where she was lying. He rubbed her back while she stared at the hole in the sky that offered so much wonder. Before long the moon and her Daddy's touch had lured her into dreamland. When she died, the moon became a nagging reminder for Keith of the terrible things that happen to innocent people. But then, Isabella was born.

Isabella's favorite book was 'I Took the Moon for a Walk.' He and Stacey had read it so many times that Isabella knew all the words, and for a while would read it upside down and backwards. At first they tried to place it in her hands properly, but she would say, "Isabella wants to do it this way." They surmised she was bored with reading it the traditional way and developed a method to entertain herself. They ignored her upside down and backward method and, eventually, she switched her book back around.

Isabella loved the moon so much; they asked a friend of Stacey's to paint it on her bedroom wall with a quote from her favorite book. "We raced for the swings where I kicked my feet high and imagined the moon had just asked me to fly."

Every reference she made to the moon brought Samantha back to Keith. It was like this child God had blessed him with was doing something no one else had been able to do, heal him.

So many paths crossed for two little sisters who had never met. He often thought, if they could travel in time and switch places for a day almost no one would tell the difference. Fifteen years after the death of one, the other had come to patch the wounds and help Keith see beauty again. But now, with Isabella at the callow age of two, a new battle had arrived to open fresh wounds.

Inside, he climbed the stairs and stepped hesitantly into Isabella's room. Tucked against the far corner was the empty crib; within he could see a child racked with fever and suffering what must have felt like the coming of the end. The floor beneath where he stole rest during her more quiet moments was now an exhibition of gifts. Toys and stuffed animals all arranged neatly and still in their packaging, all sentiments from well-wishers sent to the hospital and later transferred here in mass. The trigger of the infection was still unknown, and Isabella's immune response was still in question. She had to remain in a clean room where there was no place for such prizes.

Keith dropped into the rocking chair and closed his tired eyes. Isabella's face appeared before the curtain of his lids. His hand rose slowly and clutched the air in front of his face.

Rising from the seat, he made his way into the bedroom. On the lower shelf of a night stand was a collection of over a dozen homemade DVD videos. Stacey had taken great pride in every single accomplishment Isabella had made, all of which was well documented in picture after picture. Eighteen discs were crammed to their limit with still photos of birthdays and dress ups. There were also a host of DVD's capturing Christmases and vacations. Stacey had always wanted lots of memories captured, so Keith's job was to keep batteries charged, cameras ready and empty discs at hand for the slightest event that smacked of cute. Keith reveled in his directorial abilities in toddler movie making. He and Isabella had created a slew of magic moments in baby cinema that had brought mothers and grandmothers to tears on more than one occasion.

He slid the first DVD into the player and grabbed the remote. Soon he had compressed two years of happy exploits into a two hour television event. And with each fleeting image he realized that he had watched her grow without ever, truly, opening his eyes.

The telephone rang.

February 21, 2005

Research. It was becoming a large part of Stacey's world as she tried to digest everything she could find about her daughter's condition. These days she divided her time between keeping her vigil at Isabella's bedside and scrolling through documents at the CHP library computer. Here, she also kept pace with translating the latest information to the outside world via Isabella's Care Page; information she gathered from the doctors and the internet.

On the Wikipedia website she found a brief description of Cyclic Neutropenia.

> *Neutropenia is a hematological disorder characterized by an abnormally low number of neutrophil granulocytes (a type of white blood cell). Neutrophils usually make up 50-70% of circulating white blood cells and serve as the primary defense against infections by destroying bacteria in the blood. Hence, patients with Neutropenia are more susceptible to bacteria infections and without prompt medical attention, the condition may become life-threatening.*

Being ill and under a doctor's care can be frightening enough without the completely new vocabulary of seemingly esoteric words. Within that feeling of helplessness, our defenses tell us to fight or flee; and a mother who is trying to protect her daughter will almost always choose to fight. This is a battle where knowledge is power and understanding is the ultimate weapon.

Breaking the words down into simple terms makes them easier to understand. Each cell is like an egg. The yolk of the cell would be the nucleus and the egg white would be referred to as the cytoplasm. The word *granulocytes* simply means that the cytoplasm (or egg white) of the white blood cell is granular. The word *neutrophil* means the cell can be easily stained by neutral dyes for identification. *Neutro* refers to the neutrophil white blood cells and—*penia* means a lack or deficiency. Neutropenia in its simplest terms means a lack or deficiency of white blood cells.

Isabella's case provides three primary challenges. First and foremost was saving her life by eradicating the infection. The second challenge was to cover the exposed areas of the body with skin grafts and other means. And last was to determine how the infection occurred when the bacteria that attacked Isabella was so prevalent in our everyday lives.

Typically our bodies easily fight off these types of infections. The doctors began running blood tests for Cyclic Neutropenia, a disorder where the patient's white blood cells dip during regular intervals which exposes the patient to infection while their immune system is weak.

When testing for Neutropenia, an absolute neutrophil count or ANC scale was used based on the number of neutrophil white blood cells in a micro liter of blood. If the count dipped below

2,000 a Neutropenia condition existed with a slight chance of infection. From there the condition was listed as mild (a count of less than 1,500), moderate (a count of less than 1,000) or severe (a count of less than 500). There were several different types of Neutropenia. The doctors focused their attention on Cyclic Neutropenia based on previous skin infections. This form of the disorder was inherited; passed down from parents to children but not always becoming symptomatic for each generation. If Isabella tested positive, somewhere in Keith or Stacey's family history a distant relative suffered from the same dangerous vulnerability to infections.

Further on, the Wikipedia entry read:

> **Cyclic Neutropenia**—*tends to occur every three weeks and lasting three to six days at a time due to changing rates of cell production by the bone marrow. It is often present among several members of the same family. Cyclic Neutropenia is also the result of autosomal dominantly inherited mutations in ELA2, the gene encoding neutrophil elastase.*

ELA2 is a gene in cell DNA which controls neutrophil elastase—which is an enzyme that controls the breakdown of elastin—which is one of a group of proteins known as albumins—which is a water soluble protein that coagulates or turns from a liquid to a solid when heated—an example of which is egg whites.

The neutrophil white blood cells that have mutations in the ELA2 gene have short life spans, thus the cells never properly display in the bone marrow where they're created. Approximately every 21 days the cycle repeats itself; dropping from normal levels of neutrophil white cells to low levels.

During periods of low neutrophil, the body was like a country without homeland security wondering where the bacterial terrorists would strike with attacks from germs outside the body as well as those that had already set up cells inside. Threatened with a vast array of symptoms and intended targets, the body was susceptible to fever, oral ulcers, swollen lymph nodes, ear infections, skin infections, pneumonia, intestinal infections, blood infections, sinus infections, headaches, bone pain and tooth infections.

Primarily, the diagnosis involved a series of blood tests; however a biopsy of the bone marrow was sometimes required. The only successful treatment of Cyclic Neutropenia was a hormone known as Granulocyte Colony-Stimulating Factor (or G-CSF). A simple injection of G-CSF under the skin stimulates the bone marrow to produce and release granulocytes and stem cells. Though disquieting, the disorder could be treated. But the primary and secondary challenges of stopping the infection and covering the exposed tissue still lie ahead, and after what Stacey had witnessed the previous day—she knew Isabella remained in a dense crisis.

Two nurses had been rolling Isabella over onto her side and had exposed the dressing on her back to Stacey's view. And while the wound was not visible, the indentation from the missing tissue was all too apparent. So much of her little back was gone.

Stacey had desperately wanted to rush from the room, to scream out about the injustice of what had happened to her child, but she had to remain strong. She had to remain vigilant. There was no one else now. Her sister, Tanga could provide the love and support she needed to continue the fight while Keith was away, but the fight was hers. She could not expect Tanga to stand in her place, although she knew Tanga was willing to help in any way. Tanga's presence alone kept her grounded. She was Stacey's sole sibling, and while at times they had their differences—their defense of each other could never be denied. Stacey drew in a deep breath.

The evening had brought her little rest. The struggle to keep Isabella comfortably sedated was ever changing. They stopped one of the sedatives which can cause hallucinations and increased the methadone. This morning, she seemed to be resting adequately with no signs of agitation.

With her research drawn to a close, she made an entry on the "Telly" Care Page.

February 21, 2005 at 09:33 AM EST

Isabella is doing ok from the surgeon's standpoint. Her muscle looks good, and she will go back to surgery Wed. to have her back dressing changed. Our main goal now is to keep her comfortable, which is a challenge considering all the things she has going on. She will remain on the ventilator for a long time, until they get done with the bulk of her surgery. And then our long-term goal is to move her to some kind of rehab facility. Keep her in your thoughts. Love Keith and Stacey. If you click on "photo gallery" to the left of the screen, you can see pictures of her!

She closed the open file windows on the computer and exited the library. The sterile smell was slightly stronger in the hallway, but Stacey no longer noticed; she had grown accustomed to the blend of acrid aromas. She passed the now open doors of the sleep rooms, displaying their cheery-intended murals of maniacal torment. The previous night's occupants had already abandoned their respite to reconvene the security council of their child's healthcare.

Back in the waiting area outside the PICU, she found an unoccupied room and sat down in one of the chairs. Blank walls. Empty seats. Silence. She sighed. Tanga had gone back to her hotel room to freshen-up for the day ahead.

How much time had passed; 30 minutes? 60 minutes? No probably five. Her jaw tightened and her lips thinned. She rose from the seat and directed herself back through the twisting hallway to the entrance to the PICU. As she scrubbed her hands, she watched as the attending nurse checked the IV site on Isabella's shoulder. Isabella's foot twitched against the pressure of the nurse's fingertips. Even from this vantage point, Stacey could see the depression of the skin. Stacey's chestnut eyes smoldered. That nurse was about to receive one of many spirited lashings as Stacey aligned herself on a course that would rip through many of the staff members and lead to an ultimate showdown.

February 22, 2005

Stacey made her way around the arc of ailing children. After so many trips through, she thought it would get easier. It didn't. The air in the PICU was thick enough to swallow you, or at the very least, gnaw away at your insides until you possessed a hole that could never be refilled. Never in a place of such nurturing care, could one feel a greater sense of injustice. If so inclined, you could think of hundreds of people who would be more suited to take the place of these innocent souls; warmongers, pedophiles, murderers—all seemed far more deserving of the fate that seemed to hover above each bed like a black film, clouding the future of the child nestled within the hospital sheets; the dark of suffering as darkened by merciless uncertainty.

Today, a visual harbinger was added to the paste atmosphere; the curtains on one of the cubicles had been drawn shut. Beneath the curtain she could see needles, sterilized packaging now emptied of their cargo, silvery instruments that appeared too valuable to be discarded, and were all scattered about the floor. Tennis shoes, capped by cuffs of pale green scrubs, shifted on the tile with an occasional squeak. Nothing good could be happening within that obscuring veil.

At the entrance to Isabella's room, Stacey paused and glanced back at the drawn curtain before continuing into the anteroom. She scrubbed her hands, then added a gown and mask to her attire. She paused at the door, catching her breath. Being the interminable watchdog over Isabella's care over the past few days was taking its toll on her psyche. Tomorrow evening, Keith would arrive from Williamsport, allowing her a much needed reprieve.

A nurse was perched on a stool next to the mobile BP stand. She smiled as she entered.

"What's happening at the enclosed cubicle? Isn't that the three-year-old jaundice girl?" Stacey asked.

"The little girl with kidney failure," the nurse responded. "She's had some setbacks. They've contacted the family and have decided to unhook her life support."

"My God, she's so young." Stacey said.

"They're moving her to a private room. The parents are already on their way."

Later, Stacey would learn that the parents said their goodbyes and departed before it was over. The toddler would die alone.

February 23, 2005

February 23, 2005 at 04:32 PM EST

Isabella went back to surgery today to have her dressings changed and they had to remove a little bit more tissue on her back which was dead. We will know in a couple days whether there is still bacteria present or not. She did well through her surgery, but she is very drowsy and they had to increase settings on her ventilator, but overall she did well. The plan is to change her dressing every 5 days on her leg and back. Then, they may put a vacuum dressing on her leg in 10 days. Then, hopefully, they can start some kind of skin graft to her back first; then her leg. The waiting is the hardest part, and seeing her wounds. I saw the back dressing for the first time today and that was hard because so much of her little back is gone. But I know with time it will heal. This is probably the most difficult time in our lives we, hopefully, will ever face. We didn't know how much suffering and sadness there was until we were in a place like this children's hospital. One little girl has passed away since we have been here. Only 3 years old; and it is difficult to get to know the parents of many others that may not survive. Continue praying for Isabella. Your inspirational messages you send are very helpful. Love Keith and Stacey.

The headlights illuminated only that which needed to be seen. On the fringes, the rest of the night was under assault from all forms of halogens, fluorescents and neon. It was American commerce at its finest. There was a mattress selling beds, a tiger marketing gasoline, a Chihuahua peddling tacos, a Beagle bartering shoes, some Clydesdales vending alcohol and a clown hawking burgers. Though on this particular stretch of road, once you pass the drive-thru strip joint, you realize mercantilism has achieved an all-time low.

The Subaru wagon glided toward its destination with only a handful of miles remaining. *Thank God, for that*. Behind the wheel, Keith had donned his eyeglasses—something he reserved for driving purposes, although his nearsightedness warranted fulltime wear.

A little more than seventy-two hours had expired since he left Pittsburgh, and now he was on his way back. He was eager to see Isabella; and then there was Stacey. He was eager to see her, of course, but also apprehensive. Keith was fretful that three days of braving the stress of overseeing Isabella's care without his help had taken its toll on Stacey. And considering most people tend to release their tension on the ones they love, Keith was bracing himself for the possibility of becoming a human stress ball.

The phone calls started on his first night in Williamsport. He was watching home videos when the telephone rang. It was Stacey's sister, Tanga. 'You should be here,' she had said. 'This is too much to put on Stacey,' she had said.

When Stacey took the receiver, she had echoed Tanga's sentiments.

"Look, I can't ask Jim to hold my job," Keith had told her. "I know how this business works. Jim's going to have to replace me; he won't have a choice. So you need to decide what you want. If you want me there, fine, I'll quit my job and take care of Isabella. Just don't start yelling at me when this is all said and done, and I have to take on a job at a lower pay."

Keith pulled his car into the parking deck, selected a stall and turned off the engine. The low concrete ceiling seemed to be pressed down on top of him; all around him, as it attempted to swallow what remained of his optimism. Perhaps he and his family would become locked inside the belly of this beast for all eternity. He knew this was a good place. A place of rescue and care, but with Isabella's condition it was difficult to view it in a positive light. In fact it seemed as though light never found its way into this hospital and it would remain that way until Isabella was healthy enough to be released.

He climbed from his car and made the trek up to his daughter's room. When he arrived at Isabella's bedside he walked past Stacey, leaned down to Isabella, kissed her cheek and whispered Daddy's arrival. "How's Isabella been doing?" he asked.

"Her vitals are a little elevated," the nurse replied.

"Are you going to up her pain medication?" Stacey asked.

"Well, I think it's a little early for . . ."

"I think her pain doctor should make that judgment. He should at least be told."

"It's just a slight . . ."

A padded thump resounded in the room followed quickly by a second, and then a third. For a fourth time Isabella's good leg raised several inches off the bed and drove the heel of the foot down hard upon the mattress. Soon her leg was driving like a piston, smacking the soft tissue of her heel down with impressive force repetitively.

"What's going on?" Keith felt a swell of panic rise upon his skin.

"Isabella—" the nurse rose to her feet and approached the bed. Another strike of her heel came with a thud.

"Is she having a seizure or is she just waking up?" Keith asked. Thud.

"Isabella, you need to calm down," the nurse continued. Thud.

"I think we better get a doctor." Stacey spat. Thud.

February 24, 2005

"Hello." Stacey hadn't recognized the telephone number on her cell phone, but it was the same area code as her hometown.

"Hi, is this Mrs. Cole?"

"Yes."

"Mrs. Cole, this is Heather from Geisinger Medical Center. We need to get your permission to transfer Isabella back to our facility."

"I can't do that. She's not stable enough to be moved right now. I don't know anything about a transfer."

In addition to boasting one of the largest medical centers on the east coast, Geisinger Health System included a children's hospital, a drug and alcohol rehab center, multiple community practices and a growing Health Maintenance Organization, more commonly recognized by its initialism of HMO.

Long before the negative headlines and the national debate over the quality of care and the restraints they place on personal choice, the framework of today's HMO began taking shape. While most of the world's focus seemed to be on finding a faster means of transportation; both on land and in the air, a savvy clinic in Tacoma, Washington offered preferred healthcare to a group of lumber mill entrepreneurs and their workers for a standardized fee of fifty cents per month for each employee. After all, if the world was destined to move faster, we would need to be healthy enough to survive when we hit that windshield at sixty-five miles per hour.

Later, in 1929, the first true HMO was presented by the Ross-Loos Medical Group in Los Angeles. They started out with the best of intentions, people who treat the sick and injured were insuring treatment to the people they treat at low rates and an emphasis on preventative care; thus the term Health Maintenance Organization was begat.

In 1972, as the number of HMO's began to decline and a leader in the field began meeting with the U.S. Department of Health and Human Services to develop the Health Maintenance Organization Act of 1973 which provided grants and loans for new and struggling HMOs, loosened state restrictions on the organizations and made a federal mandate that a certified HMO be presented as an option alongside the tradition indemnity insurance, Geisinger Health Plan began a pilot program with the employees of the Medical Center and the residents in the surrounding area. At last in 1985, the not-for-profit HMO called Geisinger Health Plan was launched and eventually added Isabella Cole to its list of insured. But in the battle to provide low-cost healthcare by imploring preventative maintenance and controlling the frivolous expenditures of treatments deemed experimental and elective services such as plastic surgery, the special needs of some patients were bound to be inadvertently trimmed from the budget which has ignited a debate whether some decisions regarding testing and treatment are more "cost-driven" than "care-driven."

Stacey finished the conversation without committing to the transfer. The call had interrupted her lunch and now she was in no mood to finish her meal. She made her way back to the PICU to find out why Isabella's doctors had failed to mention a transfer.

"Geisinger has not contacted us about a transfer," Dr. Fulmer's jaw was firm and her gaze resounding. "This is not the first time this has happened. Isabella cannot be moved right now. Don't worry; you let us handle this. We will contact Geisinger and get this cleared up."

Later, after the staff's initial conversation with Geisinger, photos were taken of Isabella's wounds and emailed to Geisinger. Once those images were received Geisinger withdrew their attempt at the transfer. The battle for Isabella's care was not over; a similar problem would arise, however this time, the HMO world of "good intentions" would be stifled by an organization of good results.

Thursday was think-tank luncheon; a gathering of attending physicians, junior and senior fellows, residents and oftentimes professors, all joining together to create a round table discussion of the varying cases currently admitted into CHP. It was a culinary consultation where the main entrees consisted of steam-cooked fish, broiled, unseasoned chicken and *The Physician's Desk Reference*. The dessert menu may include a lemon poppy seed tart made with non-fat milk, a low-cal chocolate soufflé or *The New England Journal of Medicine* topped with *The Annals of Internal Medicine*.

One at a time, the diagnosis, treatment and observations of each patient were laid out to the makeshift committee. Then some of the best minds in critical care offered suggestions and theories which were debated with books and journals as their substantiations. Dishes were pushed aside in favor of rustic covered text books or crinkled magazines as the table was transformed into a lectern. Every doctor was welcomed to comment and every case received attention. Among them was Isabella Cole, a two-year-old female admitted with necrotizing fascitis. The child had been heavily sedated for nearly two weeks utilizing a regimen of Fentanyl, a powerful relative of the painkiller morphine, and a rotation of Ativan and Versed, both anti-anxiety and amnesia medications; with occasional doses of morphine, ketamine and Haldol.

Through a building tolerance and moderate increases to the Fentanyl, she was receiving the incredibly high dosage of 20 mcg/kg/hr (or 20 micrograms per kilogram of body weight per hour) with boluses of 200 mcg/kg/hr (a single dose of which, when administered to a full grown adult, was capable of arresting respiration and causing death). With Fentanyl in her bloodstream, Isabella's ventilator was direly important. But the child's body could not be expected to continue to absorb such massive quantities of narcotics, and her semi-conscious state left her no reaction to the outside world, making it difficult to assess the progress of her condition.

The greatest hurdle now would be addiction. Not in the psychological sense; infants and toddlers preferred to feel normal as opposed to "high." They are free-spaced 1,000 terabyte computers, hardwired with Microlife Windows Instinct, eager to download as much information as possible. Drugs only dampen the process. This addiction would come in the physiological sense. Her body had grown accustomed to the drugs and it would not be easy to withdraw them without a struggle.

It was decided the best course of action was rapid detoxification.

They would begin to wean her off Fentanyl while replacing it with dexmedetomadine, which provides sedation and muscle relaxation by blocking the chemicals in the brain responsible for physiological changes at times of stress such as mental alertness, muscle tension, increased heart rate, and the elevation of blood sugar for energy. The drug also acts as an anti-anxiety medication and reduces pain without the loss of consciousness. More importantly, dexmedetomadine was non-addictive.

In conjunction, Methadone would be used to assist in pain relief while keeping the body from experiencing severe withdrawal symptoms. The effects of Methadone were long acting which also made it ideal for what would be an extensive healing process.

Clonidine, another long acting drug in the same category as dexmedetomadine, would be introduced for long term treatment.

Ativan and Versed would be continued while she was supported by the ventilator. And while they were effective in reducing her anxiety and providing a form of memory block, these drugs could be addictive and would have to be weaned down once she was breathing on her own.

And for her occasional periods of higher stress, a sedative known as Chloral Hydrate would be used. Like a stiff alcoholic drink, Chloral Hydrate would help take the edge off of her discomfort. And like the aforementioned pharmaceuticals, this drug was also non-addictive, at least not to a two-year-old child.

Unloading the medicine chest on a tiny girl that weighed all of 17 kilograms (approximately 37 pounds) seemed extreme. But her condition was extreme. Dire. And without similar cases to reference, it was like negotiating a mine field while blindfolded. No matter how calculated their moves, the doctors knew she would need constant surveillance with quick reaction to any unwanted changes. This would be a cruel game of chance with a life hanging in the balance.

Despite all of the training and experience present at the table of the Consuming Consultive Council, in a science which relied on the standards of body mechanics and diagnostics, logical predictability collided far too often with happenstance. Medicine was still an art. Instead of headdresses, they wore lab coats. Instead of face-paint, they donned expressions of tranquil assertiveness which was too often mistaken for apathy. And instead of snake root, they prescribed dexmedetomadine.

February 24, 2005

February 24, 2005 at 01:23 PM EST

Hello everyone, again thank you for all your messages, prayers, thoughts that have helped to get us this far. Isabella had a rough night with trying to keep her sedated, but they stopped one of her sedative drugs that can cause hallucinations, and increased her methadone which seems to have helped. She is much calmer now and appears more comfortable. We started to talk about weaning her off the ventilator, but there is some concern over her leg wound being open and her response to pain, so we are not sure when this will happen. The immunologist spoke to us again about the cause of this bacteria, and all her immunology testing has come back normal so far. They are still trying to rule out "Cyclic Neutropenia" which is when neutrophilis dip about every 3 weeks or so, which predisposes her to infection. They are doing DNA/ genetic testing to r/o this disease. The piece that doesn't fit for her is that she was an extremely healthy child, except for a couple of skin infections, which they think is the key to all of this. Anyway, if she tests + for the gene, they treat it with Neupogen, which she seems to respond to very well. Anyway, this is such a rare disease that the few number of children who have it and also have the bacteria Clostridium septicum (which is the bacteria she has) have all been found to have Cyclic Neutropenia. I will keep trying to post updates daily. I am glad that you have been sharing them with all our staff and physicians. Love Keith and Stacey

February 25, 2005

The PICU waiting rooms were isolated from the daily activity of the rest of the hospital. From the primary hallway, a solid wood door opened at the entrance of a long, more narrow passage. The first door on the right was a restroom. On the left, was a conference room followed by two respectably sized rooms furnished with a coffee table, a loveseat and several chairs. The walls were bland and windowless. Across the hall from the final waiting room, was a more spacious room that boasted a television, a couch and a small table and chair set. Beyond those makeshift living quarters, the cramped corridor came to an abrupt dead end with several telephones that had been mounted to the wall, each with its own stool.

Keith and Stacey would often see the elderly father of the boy in the wheelchair at these phones. Sometimes with the receiver held to his ear, talking quietly with whoever was on the other end of the line. Other times the man simply sat on the stool with his head bowed in silence, choosing to banish himself from the other waiting families. Keith never understood why the man elected this self-imposed ostracism, but he was saddened by the sight of the lone man sitting at the telephone, with seemingly no one to call.

When a child was in surgery, the hospital staff requested that the family of the child remain in one of the two waiting rooms. Here, the surgeons would update them on their child's condition.

Keith and Stacey were becoming all too familiar with these four walls which served to dampen the dings of the elevators; the symphony of random conversations and the occasional message over the hospital intercom system. Once again, Isabella was in surgery. Once again, they labored with the magnitude of her situation. The surgeons were sifting through the muscles of his daughter to find and remove any dead tissue. The samples would then be sent for testing to look for traces of the microbe creatures that were eating her alive.

Keith continued to wear his armor of optimism, but longed for the moment when the battle for Isabella's life would turn to victory; or at least when the scales would be tipped in her favor. He had faith in its coming, and he had his mantra; 'I will not leave this hospital without my daughter.'

To Stacey, faith was intangible. It was a warm and fuzzy thought to someone who happened to be freezing to death. At some point, someone would have to hand that person a coat.

Isabella had so much promise. She was the perfect little girl she always wanted. There was no limit to what she could have accomplished. She was beautiful. Stacey was prepared to do anything in her power to assure her daughter's success. Now, it appeared the only things in Isabella's future would be pain, wheelchairs and deformity. They were cutting away pieces of her perfect little girl. They were destroying Isabella's future and they didn't care. The doctors' success ended at her survival. They wouldn't be there to care for her after all the cutting was done.

They wouldn't be there to explain to Isabella why she can no longer sit up by herself. And if her leg would have to be amputated, where would they be? Celebrating their victory? If she lives, what kind of life will they leave her with?

February 26, 2005

Twilight; time to sleep and perhaps to dream. It is a time when the ichor of nocturnal silence coats the light deprived air and provides cloak to the artifice. The daylight shapes suddenly shift into archetypical demons; scavengers who prey upon men who are weakened by the blackened tangles of the things they cannot see.

The Scandinavians believed that a she-creature of sleep paralysis and night-terrors, called Mara, sifted through the ranks of the semi-conscious to find her next victim. It was said that she would sit on a man's chest and ride him into a phantasmagoria of his visceral fears. This beast with ragged, sharp teeth; yellow claws; bristling white hair that sprayed out from coriaceous flesh is the very root of the word "nightmare." And tonight, the Mara was on the ride.

Keith bolted upright in his bed with a sharp pain knotting his chest and stomach. Laboring for oxygen against a heavy feeling beneath his ribs, he struggled against an unseen force that attempted to keep him pinned to his mattress. Quickly undoing the buttons of his shirt, he looked down at his torso to examine the cause of his sudden anguish. The skin was black and swollen.

He had seen this before. It was the same discoloration, the same gas-taught tissue bloating, and the same prickly dread of human suffering that Keith had witnessed attacking his daughter. It was all there again, in his bedroom. The emergency room at the Williamsport hospital. The first signs of the disease. The terrified look on his daughter's face as she pleaded with him to fix the internal bleeding which spewed from the fountainhead of her lips.

He could feel the gases burbling beneath his flesh as the microbes feasted on his living tissue. He could feel the macabre sensation of decaying alive. The disease was spreading rapidly, threatening to stop his breathing and soon his life. Gas gangrene, the anaerobic, necrotizing bacteria that threatened to steal his daughter, was now attacking him. Panic pulsed through his veins and seeped from every pore until the smell of terror coated his body in a repulsive film.

Satan must have taken great satisfaction in his creation. The microbial killing-machine from hell must be the pride and glory of his biological war-chest. The eaters of the dead as well as the living.

Eyes wide and gasping for breath, Keith struggled from the bed only to balance on unsteady legs. The doorway leading from his bedroom seemed miles away. Through cloudy vision, he used anything he could put his hands on for support as he made his way to the threshold. The air in the bedroom was so thin and cold he would not have been surprised to learn that his house now rested atop the Himalayas. His mind struggled to work amidst the dizzying faint which prevailed over his senses. The only chance of survival was to make it downstairs to find help.

Rising above himself, he could see the grotesquely distended stomach, now bruised to a blackened sheen; skin taught to the point of failure. He carried it like a mother in labor, arms wrapped across his navel as he tried to keep it from bursting open. The gas and muck of puss and blood within him rose up and pinched off every breath from his laboring lungs. Down the staircase and into an empty dining room, desperation staving off the notion that he was alone, he pushed onward with a doomsday clock poised on one minute till midnight. Death would be the easy way out; avoid the tearing of the flesh and the view of his insides as they unnaturally slid out of their confines.

Slowly a reverberating wave pushed its way into his fading existence. An ominous sound wave of despair that twisted his stomach and sickened his heart until it touched upon his dream dampened eardrums.

". . . parents?"

Keith pulled himself from the depths of sleep and the mareridt that gripped him. The Mara disappeared into the night.

"What's wrong?" Stacey, who had been sleeping in front of Keith on the waiting room couch, snapped as her body jolted to a seated position.

Bewilderment was a temporary companion. As Keith regained his senses, he realized what the male nurse had asked. *Are you Isabella's parents?* Alternative circumstances would have prompted a proud and resounding, 'yes', but the air in the ICU was tense enough to be plucked like a string and no contact with the hospital staff was taken lightly.

"It's nothing to be worried about," the nurse continued. "Isabella pulled out her vent tube. She's breathing on her own, but she is struggling a little bit so the doctor wants to reinsert it."

"What?" This time Stacey's question was not spoken out of confusion; instead it bore the sharp tone of castigation. The nurse, who had seconds earlier attempted to gently awaken the couple, straightened his body quickly as he retreated from Stacey's sudden advance. The color seemed to wash out of the nurse's face, only to reappear in the reddening expression of pure frustration prevailing over Stacey. As they ran to the ICU amidst the twilight grog of 1:00 A.M., the male nurse's lame attempt to smooth the situation became a burning ember for Stacey's agitation. She knew what it took to reintubate a patient. The metal tongue depressor forced into the mouth. The tube shoved down the throat. The very act of a patient pulling out their respirator tube can cause injury. *Nothing to be worried about?!*

The tubes are equipped with an inflated bladder which prohibits any fluids or regurgitation entering the airway from the esophagus. If the bladder is pulled back through the vocal chords without proper deflation, the patient's speech can be permanently damaged.

The distraught parents and the staggered nurse, who was now wondering what twist of misfortune caused him to morph from messenger to antagonist, filed into the hallway of the PICU entrance and were asked to wait outside until the doctor completed the intubation.

To say that Stacey gets hot under the collar or she gets fired up or even that she's burning with anger is a gross miscalculation. Stacey has always been more the gasoline type when it came to the protection of her daughter. Add some compression and a little spark and you're going to get an explosion. And on this level, bringing the safety of her child into the mix was like adding nitrous oxide. Her face creased with concentrated fury as she stood in the hallway with her arms folded at her chest. The body language was simple; she had already closed herself off to any explanation they could offer.

Keith leaned against the far wall with his hands tucked away in his pockets and waited for ignition. The minutes were passing sure enough, but it seemed as though time was finding some amusement in dragging this uncomfortable situation out as long as it could.

A sigh passed between clenched teeth as Stacey hissed, "This is ridiculous. It's unacceptable."

A short time later she spat, "I've seen what happens when they re-intubate a patient. They can damage vocal chords and break teeth."

A female in her late thirties pushed her way through the door and approached Keith and Stacey. Her light brown hair was disheveled and her eyes displayed clear signs of exhaustion through her glasses, but her gaze was hard and resolute. She introduced herself as one of the staff physicians and explained that she had just re-inserted the ventilation tube.

Stacey's gaze was sharp and unrelenting. The doctor was a new face to add to the other dozen who had stood before them. But this doctor may as well have been a matador waving a red cape of incompetence. No matter what her part in this fiasco, Doctor El Bulls-Eye was now the primary target for the charge of an enraged Stacey. "How did this happen?"

"Isabella pulled her tube out . . ."

Stacey immediately launched into her next question before the first could be properly answered. "Why wasn't she restrained? Who was the nurse on duty?"

With a tightening jaw, the doctor responded, "It was Tiffany and . . ."

"What was she doing when this happened?"

"Look, I know you're upset . . ."

Stacey's face flushed to a deeper shade of red, "This is unacceptable, of course I'm upset. What is being done to make sure this doesn't happen again?"

"It was an accident . . ."

"She's a two-year-old!" Combustion. Spark. Ignition. "Someone should be watching her constantly!"

"Look!" In the midst of an overnight double shift, patience is no longer a virtue; it's an illusion. "You have all my nurses back there on pins and needles. I'm not going to have you carrying on. It happened, and I reinserted the breathing tube, and Isabella is fine."

"She's not fine," arms still folded, Stacey leaned into her assault. "And she should have been restrained! I told those nurses more than once to keep her arms restrained so this doesn't happen!"

"Raising your voice and scaring nurses is not going to help your daughter." There was no tone of reason in her voice; the statement was structured to scold.

"Well, I don't need a lecture from you."

"You're obviously upset. I'm going to get some rest. I'll talk to you when you calm down." The doctor turned on her heels and stalked away.

The embattled mother reeled around and advanced toward the PICU entrance. Behind the glass, the Door Monitor pushed the release button without request to allow Stacey access.

Like a storm chaser watching a churning black mesocyclone for the next tornado to form, Keith followed in Stacey's wake, though not with the enthusiasm of an amateur meteorologist. To him, it seemed counterproductive to strike out at the same people they had entrusted with their daughter's care, and the ferocity of Stacey's methods made it uncomfortable to back her position. She seemed to lack compassion for her Pittsburgh colleagues. Lost were the 'I know how that feels' and the 'I can sympathize with your situation.' Drawn into the reluctant party with a sense of loathing, she approached the situation like a news anchor during ratings sweep, engaged in an expose of everything that is wrong with the medical field. Keith was left to follow while getting a camera-eye-view of the carnage. He himself never complained at a restaurant, knowing full well what it felt like to be on the other side of the serving table. His only tolerance was in the knowledge that he was comparing life to lunch. All-in-all apprehensive, he resigned himself to silence and to attempting to maintain peace through covert coercion.

After a thoroughly aggressive scrubbing of the hands, Stacey entered the room where Tiffany still tended to Isabella's care. A moment later, Keith entered with a wary camera eye on the wrangling in waiting.

Tiffany turned to face the parents, humid-eyed with cheeks still dampened from recent tears. Until the fragility of this moment, Tiffany had been one of the couple's favorite nurses. All that had yet to change was about to.

A common cause of nightmares among the employed is anxiety driven visions brought on by work related stress. Doctors often have dreams of unsuccessfully treating a patient who does not respond to any form of therapy. Taxi drivers have night terrors about the inability to find a simple location. Workers cursed with quotas find themselves waking up in a seated position in their bed, bodies deluged in the sweat of utter panic, hands frantically trying to accomplish an impossible number of tasks with only a few minutes left on the clock. Ask nearly every working class proletarian or any high ranking administrator subjected to deadlines and cost-cuts who have occasionally gone under the spell of the mare-rider and you will unearth stories of unresolved office issues. Ulcers have been formed, blood pressures have been permanently elevated and anti-anxiety medications have been prescribed, all in the name of earning a living.

For Tiffany with the tear stained cheeks, the approach of a furious mother would be fodder for many such nightmares.

Isabella's mother did not utter a word, but her glowering expression stung Tiffany all the same. Stacey began her routine of removing and re-tucking all of Isabella's blankets.

Having already faced the inquisition of the attending physician, Tiffany's body trembled beneath the duress. The doctor had scolded her like a mother who had caught her teenage daughter arriving home after curfew. Tiffany answered between gentle sobs, "I don't know. I only turned my back for a minute, and she got out of her restraints." Beyond her call of duty and her accountability to both staff and parents, she genuinely cared about the precious girl with the loose ebony curls. In allowing Isabella to remove the breathing tube, she had let Isabella down, and that was the hardest failure to accept. Her co-workers and superiors will look beyond the gaffe, knowing they have made similar errors themselves. The parents anger would subside or be redirected. But the disappointment she felt in herself levied by both the dogmatic Tiffany and Tiffany the sentry of Isabella will cast the heaviest stones of all. She knew full well the implications and repercussions of miscues in her profession, and she had readily accepted them. However, accepting her fallibility was proving far more difficult.

Stacey's eyes stared back at her, the irises appearing almost black as her mind appeared to be rollicking with contemptuous venom.

"I'm sorry," Tiffany's irriguous voice was slightly above a whisper.

"This had better not caused any damage to her vocal chords or her teeth," Stacey's voice had tightened to a hiss.

The air inside the room had become so palpable with tension that each breath was shallower than the last. The walls were closing rapidly around the young nurse and she needed time to compose herself before finishing her shift. As the parents busied themselves at the bedside of the patient, Tiffany with the tear stained cheeks retreated into the anteroom then out into the open air of the hallway. Never had the walls of this passage seemed so vast. As she hurried away from the isolation rooms, the nightmare folded away behind her. Tucked away inside a corner of her mind, it would be revisited many times over in the days and nights to come.

February 28, 2005 (Early Morning)

The backlight from the hallway silhouetted the pictures hanging on the window of the anteroom. It was evening. The lights inside the isolation room had been dimmed giving everything in the room a tarry tincture. The images of Isabella in happier times were now shadowed. The music of the monitors played out a quiet vibrato of clicks and beeps. A vinyl chair sat dormant in the east corner. Her parents were getting much needed rest in one of the sleep rooms.

The wall in the far west corner was decorated with cards and letters. Greetings conveyed with pastel teddy bears and balloon hearts carrying phrases like; Get Well Soon, Feel Better, Wishing You A Speedy Recovery and How Do You Mail A Hug? They were nothing Isabella could read; nothing she could see nor hear. Her body was still. Her mind autonomously undulated in a residual ocean of narcotics and soporifics; floating somewhere diverged from the body. She was being weaned off her medications, but she remained unconscious and unresponsive. If she was aware of her limbs at all, they felt heavy and gravity-pressed to the mattress as though they were being pushed down instead of pulled. The entry site of her IV needles would surely be irritated; the catheter site;

the ventilation tube inserted into her throat; the open wounds now wrapped in antimicrobial dressings.

If she could have opened her big brown eyes, she would have seen them coming; screamed out loud; recoiled in terror; shrieked in apprehension for the pain they were about to inflict; begged them to be 'all done.' Instead, she remained silent and approachable. Her lashes fanned out above her cheeks like a cherub in diligent prayer.

They approached as a synchronized team, wafting across the floor—their white leather uppers bringing no audible trace against the tile. They placed their supplies next to her on the bed and set about removing the protective coverings of her wounds. She remained composed.

Busy in their work replacing the dressings on the wounds, the pair never noticed Isabella's eyelids rolling like a sheet in a strong summer wind as REM sleep created a rush of brain activity. Suddenly, Isabella's quiescent posture changed. Her hands rose from the mattress and came together above her chest. The nurses paused, uncertain if she were having breakthrough discomfort. They watched as her hands pressed together and then pulled back apart as though they each held a magnet of fluctuating poles. The nurses exchanged puzzled glances until the piston movements became more fluid. The pink flesh of her palms made a subtle slapping sound each time they came together. The realization of what was occurring brought tender smiles to their faces. Isabella was clapping to the video in her mind.

March 3rd, 2005

Isabella's case generated interest throughout CHP, which was, by design, a teaching hospital. Every move the medical staff made was carefully noted for use in future cases. Her condition and response to treatment were unique. Notwithstanding the massive amount of narcotics and analgesics the child had absorbed. As her withdrawal headed into its eighth day, concerns arose—she was not waking up.

On the evening of Tuesday, March 1st, the anesthesiologist spent most of his shift monitoring Isabella's progress. It was rare for a child of two to be in detoxification so every opportunity to observe had to be exploited.

Bringing her to a heightened state of awareness was more of an art than a science. Her medications were carefully titrated to keep her within the boundaries of relative comfort while still advancing her withdrawal. Pain management was still a concern, so they allowed Isabella to dictate the pace. When her meds were at their peaks, she would become dazed, lethargic and weak. When at their lowest, she responded with agitation, sweating, jitters and sometimes outbursts of pure anger. At those times she seemed to be struggling with herself. There were odd movements of her arms and legs or hands. She would move her mouth as though tasting something foreign to her. She would grind her teeth. This spoke volumes of her discomfort.

Sometimes the symptoms were subtle. She could gaze off in a single direction for hours or send up a flare to the outside world with a combination of an elevation in her blood pressure and heart rate. Most times, it was impossible to comfort her; like a child who cannot wake up from a bad dream.

The kicking and screaming that accompanied Isabella's fits of agitation made the possibility of ICU Psychosis plausible. When the intense vomiting and diarrhea began, the symptoms appeared more like narcotic withdrawal. Meanwhile, high fevers and low neutrophil counts kept the physicians chasing various infections and continuing the treatment for Cyclic Neutropenia. A hematologist was added to her growing list of specialists.

When Isabella vomited, the staff would suction her mouth to avoid regurgitation from entering her lungs, possibly causing pneumonia. Isabella reacted to the suction with every ounce of fight left in her—upsetting to parents, but a good sign of recovery for the doctors. Yet, this was subconscious efforts of resistance. She would not respond to requests nor show any signs of general awareness.

Keith and Stacey continued their efforts to reach her. Nothing worked. These were times of great trepidation. Throughout her favorite videos, music and word games, they awaited her response—none came.

Today, the Wiggles brought their pop music and colorful antics to the attempt. Bouncing around on the television screen, the quartet sang "Hot Potato" as though life were nothing more than bubble gum and butterflies.

Isabella was showing notable signs of agitation. Beads of salty droplets formed on her forehead. Her body shook as if an electrical current were being applied to her spine each time she attempted to move.

One substance that was being withdrawn from her regimen was chloral hydrate, a drug that acts like alcohol in the body. Subsequently, when the drug is removed it can trigger a mild alcohol withdrawal.

In Harlem in the early 1900's they called it the 'jitterbugs.' In other cultures, they were called 'the shakes', 'the fears' or 'trembling madness.' Doctors refer to this condition as delirium tremens. The difficulty in pin pointing the cause of Isabella's distress was in the severity of the reactions. The shaking from her movements was mild enough to be considered muscle weakness resulting from immobility and body stress. One common denominator was elevations in heart rate and blood pressure; her most prevalent symptoms.

While her parents tried to calm her body, the world's greatest self-soother remained unconscious with her heart racing and her blood over-exerting itself through her vessels.

It was early afternoon and her latest doctor would soon be arriving on normal rounds. Since her arrival, Dr. Patel was her third physician, although the entire staff was aware of her case and always involved on the periphery. Each remaining like a member of a basketball squad; keeping aware of the status of the game in the chance they may be called from the bench. The attending physician was like the point guard; the primary ball-handler in charge of passing to the skilled players. The other four members of the starting team would be the surgeon, internist, diagnostic specialist and neurologist. If Michael Jordan had been a doctor, he would have been the most skilled surgeon in the world.

Isabella's first attending was a stout, balding man, her second doctor was tall and thin with thick dark hair. Dr Patel was a diminutive female who shared a common geographic origin with her two previous colleagues—they were of middle-eastern decent; a commonplace sight in rural-town America, due, in part, to the Conrad Program.

An increasing longevity of American life expectancy coupled with the aging baby boomers had placed a strain on the healthcare system. Small towns across the country found themselves in a severe doctor shortage. To entice highly qualified foreign students into remaining in America following their medical training, democratic Senator Kent Conrad sponsored a program that would ease the visa restraints of graduating medical students provided they agree to certain stipulations. Prior to the adoption of the Conrad Program in 1994, foreign students were obligated to return to their countries of origin to practice for two years before petitioning to re-enter the United States. The Conrad Program waived this requirement provided the new doctor agreed to practice for a minimum period of three years in what was called a health professional shortage area. Talented individuals from many countries were eager to practice medicine in what appeared to be a lucrative environment. India and Pakistan became primary exporters of healthcare professionals. Unfortunately, many found themselves facing unfair working conditions and pay. But as time went, new foreign residents followed previous ones into states that were more immigrant-friendly. In addition, a large portion of those participating in the program chose to remain in close proximity of the schools they attended. As a result of these factors, Pennsylvania became one of four states with a successful retention rate.

"Good afternoon." Dr. Patel unhooked the chart from the foot of the bed and reviewed the current status. The nurse remained in the background only smiling when making direct eye-contact with the parents.

Stacey cleared her throat. "She's flinching a lot."

"That's good." Dr. Patel uttered in passing as she eyed the file. "We want her more alert."

"Why are you removing the Fentanyl so quickly? We don't want her in pain."

Keith brushed his fingertips across Isabella's brow. He looked at the lashes fanning out from her lids. She was just sleeping.

"I understand that," Dr. Patel said, "but . . ."

"She's having very serious withdrawal without the Fentanyl," Stacey interrupted.

"We're replacing the Fentanyl with valium to keep her more . . ."

"Why can't you give her something else to keep her more comfortable?"

Isabella's skin was so pale, but Keith could still see enough rose in the cheeks to comfort him.

"We haven't taken it away completely. We have weaned it . . ."

Stacey said, "You don't understand. She's having tremors and fits of agitation."

"I do understand that. You need to take this one day at a time. The trembling that you are seeing is simply a result of the weakness in . . ."

"Well, why does it have to be done so fast? Isn't this putting additional stress on her system?"

A curled lock of hair had affixed itself next to Isabella's eye. Keith noticed her twitch so he carefully moved the hair aside.

Dr. Patel's face reddened. "I don't know what you want me to do. You will not allow me to wean . . ."

"Well, I don't want her suffering. You people don't seem to care about that."

Dr. Patel turned to the nurse. "Just go back up on the Fentanyl."

Keith spun to face Dr. Patel. "Wait a second! You can't just do that."

Dr. Patel's face darkened. "I don't know what you want me to do?"

"We want you to treat our daughter."

Stacey discretely removed herself from the conversation.

Dr. Patel threw her hands in the air in exasperation. "I am trying to treat your daughter."

Keith had sat idle while his wife questioned the staff's methods and reasoning, not having the knowledge to confirm or deny her allegations. She had been desperately attempting to control her daughter's care, but Stacey could be blunt and sometimes even cruel. He had soon become a stealth mediator, urging Stacey to temper her approach while placating the staff with looks of empathy and words of encouragement when she wasn't watching him. He lacked the medical training, but this situation called for logic. It was not regarding technical nuances; this was a simple matter of, "I know what is best to keep your daughter alive, but if you are going to question me—I no longer care."

"You need to step back and let me do my job." Dr. Patel urged. "This has been going on for a long time with many of the staff members and personally; I don't have time for it." Dr. Patel stormed from the room, leaving an uncomfortable nurse in her wake.

Keith glared at Stacey; the veins in his neck flaring; his face flushed crimson; his eyes wide as shotgun barrels. "I don't want her treating Isabella anymore."

Stacey remained stoic.

March 4, 2005

It was early March, but winter was not prepared to relinquish the icy grip it maintained in the east. Frost pressed its lips to the ground and bade the vegetation beneath to sleep with a hollow "Sssh." Life could not awaken beneath the influence of a dip in the jet stream known as the North Atlantic Oscillation which pushed frigid air from Greenland down south causing the evening temperatures to plummet into the teens. Snow threatened the forecast and kept thoughts of an early spring lingering in the background as thermal coats and elevated heating bills were still foremost on everyone's mind.

Inside Isabella's hospital room there were no windows to indicate the time of year. The air was still and the only threat of a storm was concealed in Stacey's growing frustration. Stacey's focus remained on Isabella's unresponsiveness despite the reduction in her pain medications. The only thing Isabella was becoming more responsive to was her pain. And to add rage to raw nerves, the medical staff were so transfixed on waking her up, they appeared apathetic to her discomfort. Her squabble with Dr. Patel was another example of that. At least this time, the odds were more in her favor.

Until yesterday's battle, Stacey had been waging the war alone. She had begun to wonder which side Keith was on. At times, he appeared more sympathetic to the staff than he was to her. She was the one with the sick child. She was the one who was watching out for their child. She was the one who should have his support. Now, Keith had finally spoken up to defend her against Dr. Patel.

Stacey yawned and rubbed her fists against her tightly shut eyes. It had been a restless night and the strain was manifesting itself in her concentration and her spirit. She felt ragged; she could only image how she looked. Her complexion was undoubtedly pale. She needed a hot shower and a little self-pampering. Her dark hair, now lifeless, was pulled back into a ponytail conjuring images of a raven with its wings tucked against its body. The hairstyle worked in concealing the need for a generous shampooing.

Keith sat in the vinyl seat next to the bed, eyes vacant, wandering deep somewhere in the tangles of thought. She drew in a deep breath and released a sigh. The room suddenly felt stale and she wanted out.

"I'm going to take a little walk," she said. Then after a pause, "Probably down to the cafeteria. Do you want anything?"

For a moment he didn't respond; at last he released himself from deliberation, seeming almost stunned by the intrusion of reality. "No, I'm good."

"I'll be back."

She exited Isabella's room, past the beds of the PICU, through the doors and out into the equally stagnant air of the hallway. When this was over, these drab walls and low ceilings would be destined to haunt her forever. As she moved past the entrance to the waiting rooms, the approach of the head nurse caught her attention.

She was prepared to offer the conventional morning greeting when she noticed the portly woman's deliberate approach. It would be the head nurse who spoke up first.

"Can I speak to you in the conference room for a moment?" Her tone was polite, but held a firm undertone.

Stacey agreed and followed with reluctance. She was not at her best this morning. Exhaustion was creeping up on her, leaving her feeling somewhat vulnerable. As she entered the conference room, she was greeted by Dr. Patel, the only other person in attendance.

"Have a seat, Stacey," the head nurse offered as she seated herself. "We wanted to speak to you about the way you're handling the staff during Isabella's care. I understand how difficult this must be for you; your daughter is very sick and I know you only want what is best for her. But you have to understand, these constant clashes you are having with her caregivers can be very counterproductive."

"Well, if I see something I don't agree with or understand; I'm going to speak up." Stacey said. Stacey fought against the flow of tears filling her eyes, but they spilled down over her cheeks as an involuntary reaction to stress.

"I don't blame you for that," the head nurse continued, "but you have to state your concerns more appropriately."

Stacey brushed her fingertips across her cheeks. "It is difficult to be in this situation."

"Even so, you need to be less critical of the doctors and the nurses caring for your daughter," Dr. Patel said. "We want what's best for her."

The head nurse cleared her throat. "Again, we can sympathize how difficult this must be on you. You're miles away from your family with all of the uncertainty of Isabella's condition. Perhaps you need to talk to someone; voice your feelings. We have counselors available if you would like."

"Look, the desperation to get my daughter better is always going to override everything else including how I treated the medical staff."

Dr. Patel leaned forward in her seat, "We are fighting hard to keep her alive and you should be grateful for that."

Stacey's attention drifted toward the door; not as a response to the sound of someone entering, but as a desire, as though her wish for reinforcements might cause an ally to materialize. The world was twirling and her mind could not keep up with it. She had been fighting a desperate battle against the same brazen army she had employed to defend her child, and no one else seemed to be aware of the danger their missteps were causing. Her frustration and exhaustion was zapping her strength. Her will to continue the struggle waned momentarily in the face of the insurmountable odds.

"I think Keith should be here," Stacey's soggy voice quivered.

"You want your husband with you?" the head nurse asked.

"Yes. I think Keith should be included on this meeting."

The head nurse rose from her seat. "Let me contact the attending nurse."

Upon returning to the room, the head nurse announced that Keith was on his way. She offered Stacey some tissues and the three sat quietly while they waited.

Discussing your daughter's treatment to keep her alive could never become mundane, but for Keith, typical came to mind. He had been alerted, by the attending nurse, of a meeting in which his presence had just been requested. There were many such meetings since their arrival.

With such a rare case, there were no road maps to follow. The physicians would have brainstorming discussions; think tanks. And with the birth of each treatment came the inevitable sales pitch to the parents. This is how we can stop the infection. This is how we can keep her vital organs from failing. This is how we can give her optimum use of her body after she recovers. The doctors were paving new roads between critical and discharge, while Keith and Stacey were the motorists trying to smile and accept every detour being thrown at them. But it was Isabella who would ultimately have to pay the tolls.

The door of the conference room opened up to a slightly different setting than he had expected. The head nurse was seated next to the female doctor, who on the previous evening had added a new twist to the word unprofessional during their altercation. Both appeared calm with quiet resolve. Stacey sat alone, looking threatened and meek. She began to fill Keith in on the direction of the discussion.

"They think we are being too hard on the staff. They said we're scarring everyone and that it's gotten out of hand."

The head nurse addressed Keith. "If you're not happy with the care that Isabella is receiving, perhaps we should transfer her to another facility."

Keith had spent most of the previous evening criticizing Dr. Patel's actions with Stacey. It was the first time since their arrival that he was prepared for an altercation and moreover, he welcomed it. "Fine by me. Let's transfer her. Maybe we should transfer her back to Geisinger." He watched the staff members as the look of resoluteness on their faces melted into stunned anxiety.

"Understand that's not what we want, Keith."

Keith's face tightened. "I'm telling you, it's what I want."

Stacey said, "No. We don't want that. Geisinger is not equipped to handle her injuries."

Stacey's comment was a bit of a blow to his assault. Last night they were an army of two. Last night they were ready to remove this doctor from medicine if necessary. He recovered after a disappointed glance at Stacey. "After what I heard last night, I'm all for a transfer. That was the most unprofessional response to a question that I have ever heard from a doctor," he said, directing his comments to the head nurse, "Is there a way Isabella can stay here and not allow this doctor to treat her."

"Keith, that's not possible." The head nurse had now gone from aggression to placatory. "Dr. Patel is the chief physician this week in the PICU."

"Is there any way to keep Isabella here and not let this doctor treat her."

Dr. Patel said, "We have to concentrate on Isabella's care and not this bickering."

Before Keith could retort, Stacey spoke up. "I'll try to be more patient with the staff. I only want the doctors to explain more things to us on a daily basis about Isabella's care and progress."

Dr. Patel softened her eyes. "I will do everything I can to make sure it happens."

Keith leaned back in his chair and rested his chin in his hand. His eyes, still smoldering, were cast to the ground.

The head nurse said, "The main objective is to create a healthy and positive environment for Isabella's care. We all want the same thing here. And despite how it may seem sometimes, we are very concerned for Isabella's comfort."

Stacey wrapped her hands in her lap. "I just want to know what's being done and why."

"We certainly can't fault you for that," the head nurse said over her shoulder as she followed Dr. Patel out of the conference room.

Keith locked onto Stacey with a frigid stare. His eyes never broke away as he followed her out of the conference room and into the hall. And when she turned to face him, he moved within a hair's breadth of her. His eyes had gone wild and vacant. His muscles knotted with the fury of a rampaging bull blocked from his target by a fence. "After all your pissing and moaning over me not sticking up for you; when I stood to defend you, you turned your tail and ran. Don't you ever complain about me not backing you; ever again."

Stacey watched as Keith stormed down the hallway, getting smaller as he moved deeper into the core of the critical care level. She wondered if their marriage would survive this ordeal; she wondered if she cared.

There were a few places of refuge for Keith and Stacey outside the hospital campus. The most common destinations were fast-food restaurants or one particular sub shop for a quick lunch. It was amazing how a sterilized cafeteria can make a Wendy's hamburger joint seem like home. They ate in relative silence. Keith attempted to launch bits of conversation, albeit with a tone of indifference. Stacey, who was too raw to accept the chatter, took a more direct charge. She limited her responses to a single word or an occasional sardonic roll of her eyes. Their relationship often resembled a highly contested stronghold in a fierce battle; someone was always on the attack.

On their way back to the hospital, Stacey drove while Keith seethed in the passenger seat. It was time to consummate their tranquil outing.

"Why do you have to act like such a bitch?" Keith snapped.

"Don't start on me. I'm not in the mood. You have no idea what this is like for me."

"No. I'm not buying that shit. There's no reason. There's never a fucking reason. No matter what's going on, you want to bust my balls." Stacey threw the car into park as Keith continued his tirade.

"You always have to look at the bad side of everything. You're always fucking negative." The driver side door flew open and Stacey fled from the car. She ran without direction. She ran without a true destination. It was not where she was running to, it was what she was running from; everything.

Keith found himself in the passenger seat of a vehicle in Pittsburgh traffic with no driver. As he scrambled behind the wheel, the traffic light turned green. He pulled the door shut and steered toward the hospital as Stacey ran ultimately toward her hotel room.

March 5, 2005

Keith thought he and Stacey had put their disagreement from a few days ago behind them, but clearly only he had decided to rest it in the past. They were seated in a brightly lit cafeteria, maybe the only place in the hospital where light was not in short supply. Both were avid weight watchers and both were merely picking at a plate of food that a rabbit would consider an appetizer.

His attempts at casual conversation were being met with a roll of her eyes or short, snide responses. He was pretty sure her puss soured more with everything he said. "Look Stacey, this isn't easy for either one of us, but if we are going to get through this we need to stick together."

Her eyes sharpened. "I don't even care about us. That is the last thing on my mind. All my focus is on Isabella."

Keith leaned across the table. "Okay, I'm focused on her, too. But that doesn't mean you have to treat me like shit."

"I can't stand it here. This place is going to drive me crazy."

It was Saturday. He was one day away from putting two hundred miles in between them, but he was not sure his patience could survive another twenty-four hours. Not without starting World War three and maybe four, five and six in the process.

He sighed and tightened his jaw, "Look, if this is the way you're going to treat me, I may as well leave a day early."

"Go ahead and go."

He rose from the table and grabbed his tray. "I'm going back to see Isabella before I leave."

As Keith walked away, Stacey was not sure if she wanted to scream at him or hear him say that he would not go back to Williamsport this week. Being in this hospital to cope with Isabella's condition made her stomach tense in knots, but doing it alone was too much for her to bare. He refused to acknowledge what she was going through. Instead of giving her credit and support, he criticized. He never recognized the sacrifice she was making. He was too focused on himself to see what she was enduring. Her chestnut eyes began to tear, another weekend alone.

Pulverized salt dusted the grey pavement as Keith steered his SUV along the four lane highway just outside of the Pittsburgh suburb. The degenerate sky was equally grey, but offered no dancing snowflakes and no hope for sunlight. His windshield smudged occasionally from the road-salt and Keith was happy he remembered to refill his washer fluid, for once. A mixture of sixties and seventies music drifted softly from the speakers in a volume that did little to demand attention. If the lyrics were brushing his mind at all, he was unaware. The hot air blowing through the vents could not completely diminish the cold that penetrated the windows, but the burning in his face and ears welcomed the slight chill. Stacey's rolling eyes and contemptuous comments were a bitter pill to swallow. The memories stuck in his throat and choked off his sense of husbandly duty.

As he drove, the daylight faded and his mind eased back into thoughts of a decent night's sleep, until his cell phone rang like a quick-triggered alarm clock. He recognized the area code as a Pittsburgh number.

"Hello."

"Hello is this Keith."

"Yes."

"This is Brittany one of the nurses from CHP. Everything is fine. I was just calling to tell you some good news. Dr. Fulmer was in and removed the ventilator from Isabella and she is now breathing on her own."

"Oh, my God thank you." His eyes instantly became saturated. "That's great news. Thank you so much for calling me."

He shut his phone and signaled a turn at the next intersection. Any thoughts of a comfortable bed had now been trumped by the sight of seeing his daughter's lungs expanding under their own power. He entered onto the next road and found a place to turn around, redirecting his SUV westward. This was the best news he had heard since her arrival. He grabbed his phone and found Nanny in his address book. This was a phone call he would be happy to make.

March 06, 2005

March 06, 2005 at 10:14 PM EST

Hooray! On Saturday at 4:00 pm Isabella began breathing on her own again without the use of a ventilator. This is a huge step!! She is still going through withdrawal or agitation, but now the breathing tube is out. She hasn't focused her eyes on us yet and it may take a few days for that to happen. The wound vac dressing is now on her leg as well. It is very hard to look at the way her little body has been ravaged by this infection but she is a living, breathing miracle. Stacey and I had a talk with one of the doctors and she told us the 11 doctors in ICU met on the day Isabella arrived to discuss her case and the general feeling was she would not live through this. They didn't know our Isabella!! Please pray for continued healing and full use of her leg and torso. She has been through more than any child should have to endure, so I ask you all to pray for our baby every day. Love Keith & Stacey

March 17, 2005

It had been a week of small accomplishments and small setbacks, but Isabella was still, in large part, locked in another world. She was still having hallucinations and psychotic episodes, and doctors were continuing to try varying drugs to bring her out of it; including a medication called Haldol that had her sleeping during the day and awake at night. Bouts of diarrhea had forced the medical team to perform a surgical procedure to remove the vacuum pack on her leg, clean the wound and reattach the pack dressing. Her feeding tube had been taken out and reinserted to ease the stress on her bowels. Technicians ran an EEG to test for seizure activity; none was found. Now a CAT scan was being considered to check for abnormalities that could be responsible for Isabella's state of unconsciousness. She was Sleeping Beauty without a prince.

She had been moved to another section of the PICU, this time on the seventh floor; a room with a view in purgatory. The hospital staff and her parents continued their attempts to draw her out of her private world and back into theirs. And while she remained predominantly in a transient state, they received occasional signals that the Isabella they knew was somewhere deep inside the other realm attempting to establish contact with them. Her eyes were opening more and she would often stare into space and giggle. Keith and Stacey played her favorite DVDs to reach her. During one viewing of The Wiggles, she tried to mouth the words of their songs until tears of frustration trickled down her cheeks. She had previously refused her pacifier, which under normal circumstances would make a parent happy. In this case, the staff was trying to connect her with the life she had known before the terms necrotizing fasciitis or Clostridium septicum (the actual microbe that had attacked her) ever entered her parent's vocabulary. On March 10th, exactly one month since the onset of her infection, Isabella accepted her pacifier for a brief period. The greatest moment came on the previous Sunday, when Keith asked Isabella to clap her hands and she obliged. Her parents and the staff members in the room broke into applause. During the same play session, she maneuvered her fingers together to finish the rhyme of "The Itsy Bitsy Spider."

Now, her room was quiet. Keith was in route from Williamsport after finishing his condensed work schedule. Stacey was bidding her time, watching over her little girl. The neurologist visited an hour ago and had scheduled a hearing test for the next day. He had also spoke about an MRI, but said the test could wait. Since then, the staff placed Isabella in something resembling a car seat in hopes that being upright may serve to arouse her. Isabella was now sleeping in the upright position. *Queen Isabella on her throne,* Stacey thought, *snoring away.* She survived the initial infection and was breathing on her own; now, all she needed to do was wake up

Keith steered into the parking garage and claimed an empty stall. In the passenger seat, Nanny lifted her purse onto her lap as she prepared to step from the car. A cold, damp chill emanated from the concrete structure, but thankfully, the walls kept the winds at bay. Keith led the way through the glass, double doors and to the elevator. He had spent the usual three days in Williamsport and was anxious to see if Isabella had made any progress. The reports he had received from Stacey sounded disparagingly like more of the same with few signs Isabella had become cognizant. He had recently been reminded, "Don't be afraid; only believe." The thought put an optimistic bounce in his step as the elevator delivered them to the 7th floor. He was armed with one of Isabella's favorite CDs; a Christmas disc her cousin, Courtney, gave to her. It was one of those personalized compilations where the child's name is strategically placed in each song. Three months ago, when Isabella could walk, dance and twirl, Keith and Stacey would dance with her to one of the songs, occasionally waving her 'Blanky' over her face to make her giggle. He was hoping it would spark a reaction; something to say that all of the prayer and positive thinking would offer him some semblance of a happy ending.

From prayer chains to fund raisers, the outpouring of support was inspirational. A simpatico circle had formed with Isabella at the epicenter. Those accustomed to daily prayer implored God's mercy for a little girl they never met while others found a renewed dedication to daily invocation on her behalf. Messages from Web users swelled Isabella's care page. The floor of the Williamsport Hospital where Stacey worked had created a makeshift collage of photos and well-wishes on a window behind the nurses' station.

Stacey's colleagues banded together to raise funds for the medical expenses. Gifts and cards arrived for Isabella almost daily; including an Easter basket from The Wiggles with DVDs, CDs and stuffed animals.

She had survived the initial attack, but many questions remained. Would she regain full use of her body? If she indeed had Cyclic Neutropenia, how great was the chance of her contracting another infection? With all of the medication and physical shock, was her mind functioning normally?

Stacey had informed Keith that her hearing test showed no abnormalities. She had finished a course of antibiotic from a bowel infection which seemed to ease her diarrhea. Physical therapists had begun to work with her effected leg; bending and straightening it to help her range of motion. However, the struggle with her withdrawal continued.

Keith and Nanny entered Isabella's room to find her propped up in bed and alert. Her eyes fixed on her Daddy as he entered, but her expression remained mostly stoic.

Keith commented fawningly, "Look at you. You're sitting up like a big girl."

Isabella stared. Nanny smiled in the background.

Keith told her how much he missed her and sang to her to elicit her attention; to no avail. He resorted to the Christmas CD. The third track, a song entitled, "Countdown 'Til Christmas," was Isabella's favorite. As the song began, a sparkle appeared in her eyes and she flashed a broad smile. Keith grabbed 'Blanky' and brushed it over Isabella's face. Her features disappeared under neatly stitched Teddy Bears. As it passed over her head to once again expose her face, her smile broadened, and her pretty, little white teeth began exposed from beneath her lips. When the song ended, Keith replayed it and again waved the blanket in front of her face. Again, Isabella displayed a toothy smile. On the fourth recital, her smile faded as she began to grow tired, but with that smile, Keith espied the Isabella he knew. Nanny fawned over her granddaughter and then her son. She was not certain if her tears of joy were for Isabella's return or the elation it provided Keith.

As the night waned, Isabella grew nauseous and vomited several times. Keith monitored with periodic visits as they x-rayed her stomach, but found nothing out of the ordinary. They stopped her feeds and allowed her stomach to rest. As the dawn rose, Keith's worn out little 'Bird' slept nestled in her father's arms. During a morning round, a nurse commented to Keith that Isabella received more cards than anyone in the hospital.

Keith kissed Isabella's forehead and stated airily, "She has a lot of people praying for her. That is what's pulling her through."

March 20, 2005

Palm Sunday; celebrated by Christians as the day the Christ made His triumphant return to Jerusalem amidst waving palm branches. Riding in on a donkey to signify His peaceful intentions, Jesus entered the city where He would be tortured and taken on the outskirts of Jerusalem to be crucified on a nearby hill known as Golgotha (translated as "a place of a skull"). On the Christian calendar, Palm Sunday is followed by Holy Thursday (the day of the Lord's Supper), Good Friday which commemorates the crucifixion and Easter Sunday (the day of the Resurrection). However there is a holy day that does not always procure the attention bestowed upon its predecessors. Before Palm Sunday is Lazarus Saturday, the day Jesus resurrected His friend, Lazarus. Each year these holy days usher in springtime, Nature's time of renewal. Life laid dormant by winter finds new life as the sun increases its preeminence on each day. It is a time of renewal, a time of resurrection and a time of awakening.

On Friday, for the first time in six weeks, Nanny was able to hold her granddaughter. Keith watched as Nanny cradled her little granddaughter in her arms like hundreds of times before. Nanny had rocked in her chair and hummed softly while Isabella slept.

Isabella's lashes fanned out over her cheeks as innocuous as any Raphael cherub. In the confines of the hospital room, day and night blended, wreaking havoc on her natural rhythm. She slept most of the day and remained awake most of the night.

At 2 a.m. Saturday, she was staring into Keith's face while he serenaded her with her favorite songs. There was no applause. She rewarded him with a smile of approval. The evening and the smile would turn out to be a prelude for what was to come.

Despite being incognizant, Isabella would respond to outside stimuli. When an object was placed in her hand, she would clutch it, though she showed no indication of knowing what the object was or interact with it in any way. If you thought she met your gaze, you would soon find her staring at something behind you or through you. Oftentimes she would react to seemingly terrifying images that only she could see. The only audible sounds she could emit were cries, grunts or screams.

This morning she clutched colorful plastic figures of Sesame Street characters; Big Bird and Elmo. Although as Keith watched over her, he noticed something had changed. She was no longer merely holding the toys against her stomach, she was moving them. Playing with them.

Keith turned to Nanny who was seated in one of the brightly colored leather chairs, "She's not just holding onto these toys. She is moving them as if she's playing." Keith leaned down closer to her face, "Hey, Bird. Are you playing with your toys?" For a moment Keith thought she flashed a smile, but her attention waned. Her wrists curled and Big Bird and Elmo became folded in the blanket that covered their matriarch.

"Do you want to listen to music?" Keith held onto her hand
as she held onto her Elmo figure, stroking her barely visible
knuckles with his thumb. "Daddy brought in more of your favorite
CDs." Keith retrieved a CD with a photo of four men whimsically
bordered with a flow of abstract paisley. From brown leather jacket
and jeans to turtleneck shirts and wool pants to double-breasted
crushed velvet suits, the attire of the four musicians signified a
time long past for the musical taste of a two-year-old. The title of
the CD blazoned in orange, green and blue pastels read *The Guess
Who Greatest Hits*. He placed the disc into the small Boom Box.
A split second later the room filled with the sound of a mournful
keyboard and a vocalist singing "These Eyes." Keith looked down
at Isabella and she curled her little dry lips into a smile. Although
the distant gaze in her eyes made it impossible to tell if she was
smiling at him or at some invisible parade of television characters
behind him. She had heard this song many times before when
her and Daddy shared music time. The memory of it may have
surfaced through the cloud of painkillers that had been dampening
her senses.

A nurse entered the room as the song raised to a crescendo
chorus. She carried a clipboard with her to list Isabella's vital signs
from the monitors surrounding the bed. As the song trumpeted to
its finale before fading away, she checked Isabella's IV sites for
signs of irritation.

Keith cued the CD to track five, a song that once always
elicited a reaction from Isabella. A lazy acoustic guitar was
plucked as the singer scatted before crooning about an "American
Woman." He then went on to spell the words, "say A . . . say
M . . ." However, even to a two-year-old who loved the alphabet

song, this was not her favorite part. As the twang of an electric guitar suddenly reigned over the speakers, the vocalist released a prominent "Uh" as though all his air had suddenly been knocked from his lungs. And with that, from her perch atop a critical care hospital bed (inspired by a sound recorded by vocalist Burton Cummings over 35 years prior) Isabella pursed her lips and sang "Hoo." It was the first audible response Keith had heard from his little girl since her emergency surgery.

For weeks now, Keith and Stacey had made it a mission to get her to respond to them. They were willing her to wake up. Finally, on Palm Sunday, the appearance on her face was changing. He could tell she was cognizant. It was a slight change, but a difference only a parent could know. "Oh, my God," Keith chuckled and began clapping for her. Nanny and the nurse joined the applause; Isabella smiled. The sound of applause had been taught to her since birth. It was the sound of praise and acceptance, and the sound of her parent's approval.

The nurse said, "I think I can safely say that is the first time a two-year-old sang The Guess Who during her recovery."

"She has always been our little performer. That is one of her favorite songs." Keith said. He swapped the current disc for a Christmas CD and cued it to "Rockin' Around the Christmas Tree" by Brenda Lee. As soon as the song began, Isabella's eyes lit up and she flashed a big grin.

Keith was moving quickly now. He had the momentum going his way, and he wasn't about to let it fade out. He turned the CD player off and set it aside. He leaned into her face until her eyes fixed on him. Determined to utilize every game of the past, he began reciting the alphabet stopping short before the letter P. Isabella had no response.

"What letter comes next?" Keith prodded. "What did daddy miss?"

Isabella tried to smile but it failed on her lips. Beneath the layer of fog, a glimmer in her eyes displayed a reminder of who she was before her illness.

"L, M, N, O . . ." Keith made another attempt.

She stared.

"Come on, Bird. What letter is next?" Keith asked.

Isabella said, "P."

There was another round of applause. Isabella flashed a smug smile. By the sound of the clapping and the smiles on the adult faces, she had done something extraordinary. Her Daddy bent down and gave her a warm kiss on the forehead. Things were fuzzy; like a dream. There was discomfort. She wasn't sure if she wanted to wake up. However, her Daddy was calling her name now.

March 27, 2005

It was 8:50 am on a warm Tuesday morning on August 13, 1963 near Onedia Pennsylvania. A coalmine owner named David Fellin became separated from the outside world in a mine collapse. Fortunately for Fellin, he was not alone, being trapped with Henry Thorne, one of his workers. Chilled by the cold air more than three hundred feet below the Earth's surface and engulfed in total darkness, Fellin and Thorne crawled around their new world of deprivation and fear for five days with no food, no light and little hope.

When the body and mind are stressed beyond their capacity to adapt, humans can experience hallucinations. The mind can attempt to ease rising fears through consolatory visions that provide us with the illusion that a way out of the situation has manifested.

When two men suddenly appeared in the anthracite vault, Fellin and Thorne asked them for a light. Without another word spoken, the two mysterious men vanished into the thinning air of the mine tunnel.

Off in the distance in this cramped cavern, a doorway appeared. A soft blue light emanated from the door frame and illuminated a single flight of marble stairs which ascended beyond the mineshaft ceiling. Thorne turned to Fellin and exclaimed, "Davy, I'm going home! I'm going alone if you don't want to come!"

In the world outside, emergency crews worked night and day to locate the missing miners and reintegrate them with the rest of humanity. On Sunday, August 18th, a six inch drill hole was successfully completed. Now, the outside world attempted to communicate with Fellin and Thorne. The rescue workers could hear the voices of the two men, however the words were indecipherable. A microphone connected to an amplifier on the surface was lowered and soon the voices of Fellin and Thorne rose up from their dark entrapment. Lights, food and water were lowered to the men through the six inch tunnel, which was eventually widened to allow the men to be lifted to safety.

At 2:42 am on Wednesday, August 27th, fourteen days after the ordeal began—David Fellin was re-birthed into the world through an 18 inch drill hole. While Fellin was being extracted from his crypt, he sang, "She'll Be Comin' Round The Mountain."

As the week started out, the outside world was able to catch glimpses of the Isabella that once shouted, "Want Daddy to swing ya!" She cast an occasional smile and clapped her hands, but in between there was vomiting and her failed attempts at speech.

On Tuesday night, Stacey entered the room to find a freshly bathed Isabella donning a little hospital gown. Isabella had been fidgeting with her blanket as Stacey approached her bedside. When she looked at her Mom, her eyes lit up and a broad smile stretched across her lips.

On Thursday night, Keith had just returned from taking Nanny to her hotel room when Isabella woke up frightened. The dread of the staff visits as they administered treatments and tests had traumatized her enough to find fear in anyone entering the room. Keith soothed her fears and then spent the next three hours watching Elmo and Wiggles videos until two in the morning.

Keith turned off the television and told her they had to get some sleep. He rested down into an arm chair near her bedside, but the only time Isabella closed her eyes was an occasional blink. She kept her eye on her Dad, making certain he kept his promise to remain at her side the entire night. Weariness and wariness can cause a child to want to reach out for any form of communication. She could have discussed the weather, but her lack of a proper window or updated weather reports seemed to have suspended that line of discussion. So, she opted for the next best line of conversation. Through all of the innocence her large brown eyes could generate in the dim light of her hospital room, Isabella whispered, "Hi."

"Hi," Keith responded quietly.

With that dialogue properly attended to, she returned to her unwavering stare until she decided on another topic.

She whispered, "Hi."

"Hi," Keith replied.

By 2:30 am she had drifted off to sleep. Keith maintained his sentinel throughout the night. Apart from a few cries of temporary distress, Isabella slept soundly as Keith watched over her. As the light dawned on the outside world Isabella's eyes opened wide and she uttered, "Dad."

Keith accepted the cue by turning on the television and inserting the next video in the Elmo-Wiggles rotation.

Later in the morning, two nurses entered the room and were greeted by a look of trepidation from their patient. As they began to change the dressing on her stomach, pain ravaged Isabella's tiny body until she shook with unwavering whimpers. Keith drew his face close to her tear stained cheeks, "Isabella, look at Daddy. Let's sing the alphabet." They sang the song twice as the work continued below her line of vision. Isabella bravely kept pace with her Dad despite the extreme discomfort she was experiencing.

When the nurses began clearing their instruments and discarded packaging, Isabella chanced a gaze down at them and her stomach. "All done?"

"We're all finished, Honey."

Keith leaned down to Isabella, "Say you drive me crazy."

Isabella motioned her fists in the air to mimic the turning of a steering wheel, "You drive me crazy." The nurses laughed, but Isabella remained stoic.

Once the nurses had vacated the room, Isabella lay clutching her 'Blankey'.

"Baby Bird, is there anything Daddy can do to make you feel more comfortable?" Keith asked.

Isabella pulled her favorite blanket up to her nose and said, "No."

Keith was amazed at her strength and quiet resolve, but he couldn't help notice the confusion pooling in her eyes. He leaned down, placed a kiss on her forehead, and stared into her eyes. "Isabella was very sick and Mom and Dad love her so much we had to take her to a hospital so she can get all the boo-boos fixed.

Then we can go home." Her lip quivered as an eruption of tears burst from her eyes. Keith held her tight while she cried and did his best to reassure her that everything was going to be alright. Isabella drained her tears and relaxed her body enough for Keith to straighten himself away from her with a smile.

"I am so proud of you."

The door behind them opened slightly and a young brown-haired nurse poked her head into the aperture. "Would it be okay if I come in to visit?"

Keith did not recognize her, but obliged the request.

"I just came down to see the little miracle girl," she said as she approached the bed. "Hi, Isabella, I've heard so much about you. You're even more beautiful than they described."

Isabella mustered a smile.

The nurse looked over at Keith, "I kept hearing the buzz about this little girl on the seventh floor who beat the odds. They told me about the seriousness of her condition."

"The mortality rate was over ninety-five percent," Keith responded.

The nurse gave an astonished shake of her head. "I know the rest of the staff on my floor was coming down during their breaks today to meet her." She returned her attention to Isabella, "You're our little miracle girl."

Isabella spent most of her afternoon on Easter Sunday, performing from her bed for the steady stream of hospital staff. She danced as best she could from her horizontal position while she belted out her favorite songs in flamboyant fashion.

Isabella had been lifted from the darkened tunnel of her medication. And while David Fellin spent fourteen days in isolation, Isabella had been submerged for over seven weeks. Perhaps during her struggle, she too was shown a lighted doorway with an ascending marble staircase. Though for reasons known only to God, it was not Isabella's time to ascend. Somewhere through time, the distant remnant voice of a coalminer sang, "She'll Be Comin' 'Round the Mountain" while a little girl danced from the stage of a hospital bed.

The critical stage of her care was now in the past. The doctors had begun to discuss where Isabella would be sent to rehab from the debilitating injuries suffered as a result of her cure. Only one question remained, would the hospital responsible for saving her life try to interfere with her rehabilitation?

March 29, 2005

A fever and a low white blood cell count was enough for Stacey to abandon her goal. She had hoped to get Isabella out of her room for awhile, but as the afternoon drummed on—that hope was beginning to fade. Another blood test would be the deciding vote though her persistent fever was leaving little chance of a hallway excursion.

Isabella rested back in her pillow; eyes closed, an occasional snort escaped her as she breathed in deep through her nose. It had been two tediously sleepless nights and she was beyond exhausted. Yesterday, her doctors had to use general anesthesia while her vacuum dressings were changed, causing her to be relatively groggy for the rest of the day; aiding in her night day confusion. Under normal conditions the dressings would be changed while she was awake, but the process would be too painful because the coverings were sticking to her skin due to rapid healing; which, of course, turned a bad thing into a not-so-bad thing. In fact, the wounds were healing so rapidly the plastic surgeon was now considering a skin graft to cover her leg in two weeks instead of the original impression of four. Furthermore, the wound vacuums allowed tissue granulation which would ultimately aid in successful skin grafting. Once the skin grafting was completed, Isabella's catheter could be removed without fear of infection from urine contaminating her wounds; although considering her fever and her dipping white cell count an infection was already present.

As a precaution, the doctors had taken several cultures to test for Clostridium bacteria, though they were nearly certain it was not the cause as the severity of her symptoms did not match such a diagnosis.

While she was not watching over her daughter, Stacey had plugged herself into the library computers, searching for a hospital capable of managing Isabella's unique rehabilitation needs. One thought was the Children's Institute of Pittsburgh; however they were ill-equipped to handle the type of wounds of which Isabella suffered. Hershey Medical Center was discussed, but currently lacked the experience to treat the severity of the wounds.

Isabella was awake and stable, but had massive tissue loss similar to that of a burn victim, with open wounds susceptible to infection and organ damage. At two-years-old, she clearly needed a hospital that specialized in pediatrics. The specialty of a pediatric burn unit was proving to be a rare combination.

And to make matters more difficult, Geisinger Health Plan had strongly suggested either Geisinger Medical Center or the Lehigh Valley Burn Center, and by strongly suggesting, they meant—they would refuse to pay Isabella's expenses at any other facility. Lehigh Valley was a reputable burn center, but was not centered on pediatrics, which provided a varying degree of difficulty. Struggling to find a hospital that would be optimal and not simply acceptable, Stacey Googled "skin grafts" and found what would prove to be the godsend of Isabella's fight to return home, a medical caregiver that not only specialized in both pediatrics and skin grafting, a facility that had experience with the very same flesh eating bacteria which caused Isabella's wounds. This was a place that could provide Isabella with the best rehabilitation care and, Geisinger Health Plan could not affect Stacey's decision; this medical center would provide Isabella's care for free. Isabella's recovery would be contingent on the alliance of a doctor and an actor over a century before she was born.

The road to Isabella's recovery began in 1870 with the origin of an organization called The Ancient Arabic Order of the Nobles of the Mystic Shrine; when two men who were part of the fraternity of Masons wanted to form a new fellowship based more upon rejoicing than ritual. Dr. Walter M. Fleming, M.D. and an actor named William J. Florence acted upon an idea once discussed by a group of Masons in Manhattan. It was to be a fraternity within a fraternity as the criterion of all members would be a degree of Masonry above Master Mason, known as the Scottish Rite or York Rite.

The two men adopted their rituals and Middle Eastern theme from a stage production Florence had attended at the invitation of an Arabian diplomat. The same stage production influenced the organization's emblem and wardrobe; and with all rites and rules set into place, they initiated each other in the summer of that same year and later simplified the name of the organization to simply, the Shriners.

Despite its Mid-Eastern roots, the Shriners were not affiliated with the religion of Islam or any other religion, although it was a requirement for each Shriner to believe in a Supreme Being.

As the Shriners began drawing members from other countries, the name evolved into Shriners International. At the turn of the nineteenth century the organization swelled to include over 50,000 members and 82 temples. They participated in many community functions and assisted the needs of others through fund raisers. They would gather at large meetings called the Imperial Session of the Shriners to discuss and vote upon the actions taken by the organization. In 1920, during the Imperial Session of the Shriners in Portland, Oregon a resolution was passed by unanimous vote to open a hospital that would provide care for children stricken with polio at no financial obligation to their families. A brief two years later Shriners International opened the doors to the first Shriners Hospital for Children in Shreveport, Louisiana. Soon the hospital's level of expertise branched into orthopedics, cerebral palsy, spinal bifida, spinal cord injuries, and later the treatment of severe burns; and as diseases evolve—the Shriners Hospitals make every attempt to keep pace in treatments.

With the debridement of Isabella's back and leg, the wounds that remained would need to be treated as those of a burn victim. The nearest Shriner's with all of the capabilities and experience was located in Cincinnati, Ohio.

Stacey approached Dr. Katie Fulmer with the proposal of transferring Isabella to the Shriner's Hospital in Cincinnati. In Dr. Fulmer's office, the pair researched the facility and completed the necessary applications.

The trip to a new city would take place on April 1, 2005. For Isabella, it was one step away from home, but even with the most loving care—it would be a place of hellish pain. For a battle weary Stacey, it would be a place where her fight for the proper critical care would reach a critical stage.

Cincinnati, Ohio

April 1, 2005

Dr Katie Fulmer seated herself at the desk inside one of the modest two-by-three-foot cubicles and lifted the telephone receiver from its cradle. She was in a small, windowless room that had been divided into six efficiency offices that she shared with three other staff members, but for the moment, she was alone with her notes and thoughts. In this insipid room, there would be little to distract her as she dictated the Inpatient Discharge Summary for Isabella Cole. Isabella was scheduled to be transferred to Shriner's Children's Hospital in Cincinnati later that morning following nearly two months of care.

She could still recall the look on her husband, John's face when he returned from assisting in the Life Flight of the toddler. He had held out little hope for the girl's survival.

When he left that night, she was reading a book entitled *Complications.* It was a revealing look at the inside of the medical field by surgeon, Dr. Atul Gawande, a subject that in many ways mirrored the tribulations of Isabella and her family. Even the title rang true, complications with Isabella's illness; complications with the pain medications; complications with how the mother perceived the quality of Isabella's care. This was nothing new,

Katie struggled with these situations on a daily basis, yet this case brought so many of them together.

In the Introduction of the book, Gawande wrote:

"It is an imperfect science, an enterprise of constantly changing knowledge, uncertain information, fallible individuals, and at the same time lives are on the line. There is science in what we do, yes, but also habit, intuition, and sometimes plain old guessing. The gap between what we know and what we aim for persists. And this gap complicates everything we do."

The recent increase in cases of necrotizing fasciitis had definitely caused the medical field to regress into the basics of sterilization, as well as exploring new options to treat the disease. Hyperbaric oxygen treatment was one of those options; a strategy Dr. Gawande had explored in the final chapter of his book.

Katie dialed the number for the medical transcription service and entered her provider code. She would dictate the details of the transfer through the telephone system and the transcription service would provide a typed version of the diction.

The final chapter of *Complications* was entitled *The Case of the Red Leg*, which told the story of a twenty-three-year-old female who contracted necrotizing fasciitis in her foot and leg. Gawande and his colleagues had treated the young woman using the same techniques the staff of CHP had enlisted to treat Isabella, first in debridement and then with oxygen bombardment.

She moved through her notes speaking into the telephone to some unseen professional who would ultimately listen to the recording and transcribe it to paper.

"Serosanguineous foul-smelling drainage from her bandages soaking through onto the bed sheets." As she spoke the words, a memory formed triggering her limbic system to relive the offensive odor. The fluid a mixture of blood and serum, the residue of clotting and antibodies, was like the smoke from guns foretelling the battle being waged inside the toddler's body.

She selected her words and descriptions carefully (although you can't always describe a mountain lion as a cuddly kitten) knowing Isabella's family may one day read the Discharge. Isabella's mother, Stacey, knew the medical field from the inside. She knew that it was not always a perfect science; and she understood the trappings of mapping out new courses of treatment. Stacey bore the burden of comprehending more than most parents would care to know, and she held the weight of trusting her daughter's care to others while still doing everything in her power to ensure her daughter's survival utilizing her own knowledge and experience. This was a struggle depicted once again by Dr. Gawande in *Complications* during a frightening ordeal he had when Hunter, his infant daughter who had been born five weeks premature, stopped breathing. A distraught Gawande and his wife rushed the child to the hospital where she was admitted for a respiratory virus. Knowing what to do, yet being too overwhelmed to make any analytical decisions, Gawande allowed fellow doctors whom he did not know decide the course of treatment; of which he would question on occasion with fervor.

Gawande had summarized the experience of making decisions on medical treatment from the perspective of a patient in the following text.

> "Just as there is an art to being a doctor, there is an art
> to being a patient. You must choose wisely when to submit

and when to assert yourself. Even when patients decide not to decide, they should still question their physicians and insist on explanations. I may have let Hunter's doctors take control, but I pressed them for a clear plan in the event that she would crash. Later, I worried that they were being too slow to feed her—she wasn't given anything to eat for more than a week, and I pestered them with questions as to why. When they took her off the oxygen monitor on her eleventh day in the hospital, I got nervous. What harm was there in keeping it on, I asked. I'm sure I was obstinate, even wrongheaded, at times. You do the best you can, taking the measure of your doctors and nurses and your own situation, trying to be neither too passive nor too pushy for your own good."

Stacey could be difficult and at times aversive, but under the same circumstances; Katie could not say her own actions would have been entirely different. It was frightening for the average parent of a sick child who may not fully understand the course and treatment of an illness, yet holding onto faith and hope can oftentimes be easier when someone is sheltered from a harsh reality. Once you bite into the fruit you must live with the knowledge of what it bears, and the taste of that knowledge can be sickening when your child's life or death is hanging on the balance of your own acquired skill-set.

Katie finished her report and cradled the receiver. She leaned back in her chair, releasing a tiny groan as she stretched out her back. She had been directly involved in most of Isabella's care, but when Isabella was extubated, Dr. Huyla Bayir, the PICU attending on service, had decided upon a special combination of drugs to aid in a rapid withdrawal from the inordinate amount of sedation medications that had so heavily flooded Isabella's consciousness. The sedatives had become a complication in determining the status of Isabella's condition. Huyla's calculated hunch helped resolve the child's physiological withdrawal, but it wasn't always comfortable for the parents to watch. Isabella had been allowed to sleep through the enormous amount of suffering her condition inflicted, so arousing her to the pain seemed nearly inhumane; and yet necessary.

The medical profession was a balancing act between life and death; between what was needed and what could be tolerated. Their job was to aid the body in healing itself without impeding that healing. At times it seemed if Katie cast her eyes downward she would see a high-wire with the ringmaster standing in the circus ring far below proudly boasting that her act was being performed without the aid of a safety net. Katie smiled at the analogy. She held a doctorate degree in tightrope walking. The stress of her occupation was palpable, but the rewards far outweighed the risks.

Off in the distance, the baleful throbbing of rotary blades concussing a grey sky announced the approach of a Life Flight chopper. Another child was being transferred into Dr. Katie Fulmer's care.

The day began with excitement, uncertainty and sadness, but now as Stacey directed her car toward Cincinnati, an unexpected calm settled over her. No sooner had the car exited from the darkness of the CHP parking garage, she felt a great relief. The longest winter of her life had finally come to end. The sun traversed the sky in a southern arc, fanning nourishing rays against the grassy, northern embankment of the highway which folded before sloping upwards where outcroppings of rocks yielded to the first vegetation of springtime. The stems stretched to the heavens ready to burst forth their flowers in a languished display of colorful fireworks of buttercups, blood root, spring beauty and a purple wild flower whose common name is aptly 'the harbinger of spring.' Further up the ridge, the tree line displayed a faint tinge of green as the new buds popped free from their barky encasements. The drive was already proving itself to be very therapeutic. Even the remnant salt streak across the windshield, now illuminated by the sunlight, was far less annoying than it would be normally. She depressed the button to release a spray of washer fluid which mostly removed it. She was out in the open air, far from the beeps and clicks of monitors and the astringent smell of a disinfected room, and Isabella was one step closer to home.

She had arrived in Pittsburgh earlier on this Friday morning, coming from Williamsport where she had gone the night before to collect some supplies they would need for their new temporary home in Cincinnati. She wanted to be present when Isabella was transferred to the airport. Also with limited seating on the plane, Stacey was charged with transferring the belongings Isabella had accumulated in CHP. The day in Pittsburgh started out with anxiety.

Stacey had been happy for the opportunity to say goodbye and to thank the staff at CHP. The goodbyes were heartfelt and the tears shed among them were sincere. While, at times, she had her disagreements with them, she knew they had provided her daughter with superior care. She shared hugs and tears with Dr. Katie as well as with Brooke and Amanda, two of her favorite nurses. Without the stress of Isabella's care looming over her judgment, her memories of the staff were mostly pleasant ones. She could sense in them the reluctance of relinquishing Isabella's treatment to another facility, although they had no doubt in the capabilities of the staff at Shriner's. She could also sense a swell of wonderment regarding Isabella's recovery. The probability of surviving had been so low and no one knew that more than the doctors who treated her wounds. Against all odds, CHP had kept Isabella alive for two months and pulled her through the most dangerous weeks of the illness. Now, it was time for the healing process to begin. Isabella's body was coming out of a cold winter of its own, ready to replenish some of the cells it had lost.

It was a five and a half hour drive, taking her across the state line into Ohio by means of Interstate 70. Two hours outside of Cincinnati she spotted a Toys R Us and decided to stop and buy Isabella some things for her new room; her home away from home and for the next stage in her ultimate return.

Keith took Isabella's little hand in his and examined her expression. There was no sign of tension in her features, no look of concern or question. The interior of the airplane appeared to have been partially gutted. Keith was certain some seats were missing, with only two remaining; one for a nurse and one for himself. Straps came up from the floor of the plane to stabilize Isabella's stretcher, conjuring up images of the military planes he had seen so often in movies. The interior was small, but the flight would be quick, far better than an extended ambulance ride for which Isabella was too unstable to risk. Keith thought of Dr. Katie, who had taken so much care to ensure Isabella's comfort. He wished she were transferring with them. The airplane began a slow roll toward the runway, soon taking to the sky and leaving another group of nightmares in his daughter's illness behind; and below.

As the plane taxied down the runway, gathering speed, Isabella stared out the window. The roar of the engine intensified as the wheels lifted from the concrete. For a brief moment it felt as though her stomach would remain on the ground, which certainly would not help her digestion in the least. Her fingertips attempted to dig into the mattress as though she were a kitten falling to the earth instead of lifting from it, until the motion of her body matched that of the airplane. Her body relaxed and a smile curled her lips. The sunlight reflected off the polished wing and infiltrated her tiny, white-plastic-lined window which granted her a limited view of a world that was passing by in blue sky and pillowy white clouds. Here there was no exhaust-stained snow, no buildings or busy streets, only the magic of flight. She stared for a while at the cottony shapes, feeling her lids growing heavy with the passing of each heavenly headrest. Her breathing became slow and rhythmic and her view of the world reduced to slits until at last succumbing to the images of her dreams.

When Isabella awoke, her Dad was speaking to her about an ambulance ride. He seemed excited about it and was trying to make her excited too. Her bed was lifted by people she had never seen before and slid into place in the back of a vehicle with walls stocked with stuff like a doctor's room. Her dad climbed in and sat down next to her bed. He was still talking about how exciting the drive in the ambulance was going to be. She was happy for him, but she didn't feel like talking. Her mind was cloudy from sleep and she wasn't sure if driving in a doctor's office was a good thing. One of the people who put her in the vehicle got in also and kept telling her she was doing great. She wasn't sure what great thing she was doing, but he seemed to be happy about it. One thing was good, he wasn't trying to stick her with sharp things or pull the wrappings off her boo-boos. When the ambulance started to move her dad smiled and said, "Here we go." Ambulances must be like a merry-go-round for him. Sure, they were going, but where?

In the heart of Midwest America lies a city whose namesake was derived from a man who went from farmer to Roman dictator and back to farmer in the miniscule span of sixteen days.

In the year 458 BC, the Roman Empire was struggling in a war with both the Aequians and the Sabines. When the leadership became thinned by the trappings of their enemies, the Senate sent a group of men to reacquire the services of an exiled politician. While plowing the fields of his farm, the man was approached and informed that the Roman Senate had just bestowed upon him the lawful dictatorial control of Rome itself. He immediately left for the city where he utilized his new power to gather an Army against the Aequians. Acting as the commander of the infantry, he led his army in a surprise attack that ended in the surrender of the Aequian leaders. With swift refrain, the war was over and the farmer, now dictator, returned to Rome.

The entire ordeal lasted only sixteen days, ending when Lucius Quinctius Cincinnatus relinquished his dictatorial power back to the Roman Senate and returned to his farm where his plow awaited him.

Such a story of selfless sacrifice for country served as a beacon of honor to the forefathers of the new world. Following the Revolutionary War, American officers founded an organization in 1783 which would assure that the government honored the promises it made to its officers. George Washington was among the members of The Society of the Cincinnati, which adopted its name from the legendary Roman farmer. In 1819, a robust village in the state of Ohio was incorporated as a city; borrowing its name from a society of American officers—the city was named Cincinnati. In that same year, the Medical College of Ohio opened its doors to begin training eager young minds in the art of medicine.

Medicine in the 19th century was still practiced door to door. Those families who were too poor to pay for the services of doctors were funneled downward to the various hospitals. Considered by most to be the market place for the undertaker, hospitals were disease festering dormitories where sick people went to die. Oftentimes patients would contract a deadlier disease than the illness for which they were admitted. No group of people suffered more than the children who, in most cases, were treated with less regard than their adult counterparts. This fact led to the founding of the Cincinnati Children's Hospital. Group medical practices began taking on a new outlook over the next few decades and facilities like the Cincinnati Children's Hospital experienced rapid growth moving from a simple two-story house into larger buildings and ultimately into a multi-level structure near the University of Cincinnati. The Medical College of Ohio moved their institute to join the University of Cincinnati in 1896, officially becoming the University of Cincinnati—College of Medicine.

The conglomerate began drawing some of the greatest minds in medicine, including an immunologist and virologist named Albert Sabin who became affiliated with the Cincinnati Children's Hospital and served as the head of Pediatric Research at the University of Cincinnati where he developed the oral polio vaccine.

In the late 1960s when the Shriners organization began exploring locations for a pediatric burn hospital, the area surrounding the University of Cincinnati seemed a logical choice. Pillared by interstates 71 and 75 the Shriners Hospital of Cincinnati is tucked in amidst the campus of the University on Burnet Avenue just east of a road dubbed Albert Sabin Way, thus granting them access to some of the greatest, young medical minds America has to offer. The academic center was an inspirational snowball that started with a not-so-simple Roman dictator-farmer. Today Isabella Cole would join the ranks of children fighting for the simple lives they remembered before they became far too acquainted with IVs and bandages. Lucius Quinctius Cincinnatus would be proud to stand with these young soldiers who fight their injuries with such tenacity and will to survive.

Two loading docks edged with yellow dock levellers were tucked inside an alcove that was cut into the base of a large, smooth block-stone and decorative concrete building with green-marbled accent tiles and arching windows. Red-lettered signs hung on the side of the building, just above the nook, reminded delivery trucks of a thirteen foot clearance restriction. Another clearance sign, this one written backwards, was posted between the two docks to remind drivers through their rearview mirrors.

The paved floor of the delivery niche was stained with a few oil slicks, but was otherwise immaculately maintained. To the left of the docks was a large open-top waste container; to the right was a set of concrete steps with steel railing that led up to an observation booth and a door which allowed access to the building's interior. Further to the left and on the façade of the building a posted sign read "Shriners Burns Institute Loading Docks."

It seemed an unlikely docking place for an ambulance to deliver a new patient to the hospital, however something within Keith's core was already telling him—this was no ordinary facility.

As the attendants wheeled Isabella's bed in through the dock opening, Keith followed the accompanying nurse up the concrete stairs and in through a steel door.

The deeper he penetrated the inner sanctum of the hospital the more uplifting the atmosphere became to his spirits. Teal, green, pink and powder blue, the colors inspired warmth and reassurance to both parent and child. There were lighting fixtures resembling lampposts, a reception desk better suited for the railing of a carousel and toy soldiers guarding doorways. A dual chair rail ran down the hallway with a pastel, aqua strip painted between. Peeking through the hall windows were the familiar faces of Dora, the Seven Dwarfs and Spiderman. Occasionally, he passed the false front of waist-high houses mounted to the walls like the set of some Alice in Wonderland movie; brightly colored with tiny doors accessible only when Alice drinks the magic tea. Keith would not have been surprised to discover the chief of staff was Gene Wilder in his portrayal of Willa Wonka.

Teal squares were added to the off-white floor like the frame of a living picture waiting for a child to lend their image; or perhaps the squares were a trap door leading to a magical world hidden somewhere beneath the hospital. Just a simple touch of imagination had transformed the entire scene from a medical facility to that of a daycare or indoor play area. Like the sun brushing away the grey clouds after a storm, a wind-swell of faith and hope began to clear away the pessimistic gloom.

The door window leading to Isabella's room had a welcome sign with a picture of Elmo. The room was almost inviting, had it not been-in fact-a hospital room, with an additional bed for parents to spend the night with their child. While the staff surrounded Isabella to record her arriving vital signs and get her settled in to her new room, a volunteer from the hospital guided Keith to the upper level of the building to the family accommodations. He and Stacey were granted a spacious room with two twin beds, a table and two chairs, a couch, a television and their own private bathroom and shower. And best of all, they didn't need to win a lottery each night to see if they can use the room; it was theirs as long as they needed.

"These types of wounds don't scare us," the stocky bespectacled man said, his full head of gray hair was combed back in a single wave. His narrow eyes shimmered with confidence behind round eyeglasses that were all but slightly large for today's fashion and his lips were small enough to appear permanently pursued. He looked comfortable in his pin-striped suit and coat that wrapped neatly about him in a display of his success and attention to detail. "We deal with them every day. You will be surprised how quickly these kids recover once you close their wounds."

Keith liked this guy already.

Dr. Richard Eckley was the Chief of Staff at Shriners Hospital for Children in Cincinnati. He was also a professor of surgery and held the title of director of the special burn care unit at University Hospital. "The Children's Hospital of Pittsburgh did a wonderful job. The wounds are nice and clean. We should only need two graft days. In examining her wounds, I wouldn't expect her to have any physical limitations after her recovery. We will use the unaffected side of her back and her left leg as the donor sites. These areas will be painful for her after the procedure, but will heal relatively quickly. We also have to graft a small area on her belly that has yet to close."

"How long will she have to stay here at the hospital?" Stacey asked.

"You can plan one day for every percentage of affected skin." Dr. Eckley explained. "Isabella has 28% of her body affected so it should be one month to a month and half."

"What about her long term recovery? Will she need to wear compression garments?"

"Yes, for about a year. It will keep the scarring to a minimum." Dr. Eckley rose to his feet. "Tomorrow we will redress her wounds and then get started first thing next week." He offered a firm handshake to both Keith and Stacey. "You folks have a good night. Try to get some rest."

"Having a bed in Isabella's room will help with that." Keith responded.

"Yeah. Sure beats curling up on a chair, huh?" A smile broadened his lips further than what Keith thought possible. "It was nice meeting you both; and Isabella too."

Keith and Stacey followed Dr. Eckley out of the conference room and began a casual stroll through the hallway. They passed a window with one of the characters from the Disney film *The Lion King* smiling at him from behind the glass. The atmosphere was bright, almost carnival in nature. Keith felt a long absent smile perched on his face. "So," he asked, "what do you think of this place?"

"It's nice." Stacey uttered.

"Don't sound too enthusiastic. I was really anxious for you to see how bright this hospital is. I think it will help Bird's spirits." *And yours,* he thought.

"I'm not going to be comfortable until we get her home and all of this is behind us."

"Well, try to relax a little, alright. Dr. Eckley said a month, so just one month and we get to take her home. We just have to accept it and take things day by day."

Stacey did not respond. She did not allow him the slightest nod of her head. Instead, she stared at the hallway as it narrowed off in the distance until it was nothing more than a tiny black square, seemingly with no end. She could almost sense the further turmoil to come and the pursuit from the hounds of a self-fulfilling prophecy.

April 3, 2005

Rusty cocked his head and stared at the little girl, attempting to get every possible angle as he read her emotions. He sensed her sadness and uncertainty, but this one had a calm resolve hidden just beneath the surface that was rarely seen. In his nearly twelve months of experience working as a psychological therapist specializing in pediatric burn patients, he was still awed by the enormous range of emotions he encountered with these children. His assistant, Tosha, was a relative newcomer to the field. Her approach seemed to be more in the gathering of physical information rather than reading the temperament of the patient. He was like that in the beginning. She would eventually temper her approach.

Rusty felt confident in his ability to reach this patient; she seemed receptive to the session. He knew his effects had limitations, but reaching the hearts of these distraught adolescents was of great importance to their overall care. His contribution to the healthcare system was discounted by some, though he knew he could at times play a pivotal role in a child's recovery. And while he could not flaunt the degree of a psychologist, he boasted a myriad of abilities that his high-brow colleague would never be able to accomplish; not the least of which was wagging his tail.

Poised in a little red wagon, Isabella eyed the long fur of the Great Pyrenees Mountain dog named Tosha as it sniffed curiously at Isabella's clothes. A tan mixed Poodle named Rusty seated inside the wagon, gazed up into Isabella's face. It was a Kodak moment captured in Polaroid pictures by a member of the Shriners Hospital staff.

Isabella was wet, her chest was congested and her face was abnormally swollen from heavy doses of steroids, but she presented a brave smile. The medical staff had assured Keith and Stacey the wheezing was nothing to be alarmed about. The puffy face was close enough to her natural features to be dismissed. The wet was something entirely new and seemed to have her restricting her movements. Earlier, they had removed her vacuum dressings in place of bandages they soaked with an anti-biotic solution every two hours. The attention of the staff and the presence of the dogs had greatly lifted her spirits. Moments like these were all Keith and Stacey had to cling to, as they attempted to forget the rest. It was a wonderful idea, tailor-made for Isabella who loved dogs as much as her parents. She was enjoying the company of the animals and enjoying the attention it was garnering her from the staff members. Keith's stomach had only begun to settle from the news he had been given earlier when the pet therapy began.

Tomorrow the surgeons would prepare her wounds for grafting, with the grafting to begin on Tuesday. That would mark a new beginning of excruciating pain for their daughter, suffering they would be forced to help inflict. They would be responsible for cleaning Isabella's wounds twice a day during her recovery.

"It will be very painful for her," the nurse had told them, "but you just have to get it done quickly and keep telling yourself it's for her own good."

'Good' would be the last word the family would use to describe what they were about to endure.

April 4, 2005

The man lay face down strapped to a small wooden table. He had known much pain in his life, yet there was great trepidation for what was about to happen. He was a willing participant for the proceedings, but great anxiety always follows in the knowing of future suffering.

Beyond the stone walls of this small room twisting vapors of heat rose from the sandy soil of India. The only portion of his body that seemed slightly cooled in the heat was his bare buttocks; that sensation left him feeling chilled and vulnerable. He had experienced worse when they carried out his sentence for the crime of adultery. On that day he wished he had closed his eyes before seeing the instrument they had used to cut off his nose; the punishment for the crime.

This time, he would not look at what was transpiring behind his back. There would be pain, but these wonderful Hindus would grant him what could only be considered a miracle. They were preparing to make him a new nose. Thousands of years would have to pass before these miracles would be performed by skilled surgeons. The Hindus were merely potters and tile makers.

He barely had time to prepare himself before the heavy wooden paddle struck with calloused ferocity on the bare skin of his buttock. His scream was blocked by tightly clenched teeth. Again and again the paddle struck until his flank became numb with pain. The skin would have to become red and swollen with the congestion of fluid before the Hindus could cut it away with razor sharp blades.

Needles of stinging pain pulsed his flesh, his breathes heaved through each wave and he could feel warm beads of blood seeping up through his pores. As the tears streamed from the corner of his eyes, he told himself once again, the end result would be worth the suffering. He would have a nose constructed from his own skin. The paddling stopped. He felt the cut of the knife.

This fictional depiction of an early skin grafting procedure was derived from written accounts of operations performed between 3,000 and 2,000 B.C. in India. By today's standards these operations appear primitive and fairly brutal, yet it was the first step in a process that has served to rescue countless patients of burns and large tissue wounds; Isabella would be the next beneficiary of the courageous people of ancient India and all who followed.

Surgeons have arrived at today's techniques on the sometimes failed and other times groundbreaking work of their predecessors. The history of skin grafting is littered with suffering; from an Italian slave whose nose was removed to replace that of his master, to a surgeon so dedicated to helping his patients he would remove small sections of his own skin to graft, to a London research group experimenting with skin grafts on a dog until they lost their less than voluntary subject when the poor pooch managed to escape and run away.

The very premise of skin grafting, though necessary, could be described as cruel to be cured. Isabella has wounds and to repair them, the surgeons will create more wounds. How do you explain to a child who has fallen and scraped their knee that Mommy and Daddy are going to take a hand file and scrap the other knee to make it all better?

There are two methods of skin graft. The first is called a full-thickness graft and involves removing all three layers of the donor skin including the epidermis (the outer most layer), the dermis and the hypodermis which is then placed over the wounds. The donor sites are stitched or stapled back together. These donor sites are far less painful and have less scarring, leaving the areas with nothing more than an incision. While skin grafting is a form of organ transplant, the suffering has not been eliminated. The chance of transplant rejection and the pain of recovery still prevail. If they perform this type on Isabella, her body may view the transplanted skin as a threat and allow her immune system to attack the newly placed skin.

With this in mind, the surgeons opted for a split-thickness graft which removes the epidermis and only a portion of the dermis, or the top layer of skin and half of the second layer. The second layer of skin contains tiny blood vessels, arteries and more importantly for Isabella, nerves which will be severed as the surgeon uses a tool called a dermatome to remove the skin. A dermatome resembles a large, electrified cheese slicer or potato peeler and brings to mind a procedure that would be better suited for the next installment in the series of theatrical films entitled "Saw." When the nerves in the skin have their receptors cut off, new receptors grow. These receptors will begin to continuously warn the body of the damage with the message of extreme pain. The message of pain will continue until the outer layer of skin begins to be replaced by the growth of new skin cells.

Scraped skin is given cute names such as road rash or strawberries, but there is nothing charming about the pain that can be experienced; especially when the scraped areas need to be cleaned and sterilized twice a day and the skin covering both the wounds and the donor sites must be stretched to prevent loss of mobility due to scarring. Most of us can remember as a child when we scraped our knee how much it hurt when we bent or extended the joint. The equation of anguish here would be to multiple that by thousands.

Once the surgeons peel off Isabella's skin using the dermatome, they will "mesh" it by placing tiny interrupted cuts throughout the section before placing it over her wounds. This will allow the skin to be stretched in order to cover a larger area as well as to help the fluid drain from her body. If the skin is not meshed then as Isabella's immune system releases fluid to repair the damage, the fluid can build up under the new skin and cause it to slide off the wounds. And while meshing is necessary to ensure the success of the graft, it will cause the skin to appear pebbled once it is completely healed. To combat this effect, Isabella will be forced to wear stiflingly tight pressure garments for many months following her surgery.

April 5, 2005

Stacey was seated in a large, leather-bound burgundy chair while Keith was seated on a wall-unit bed which also served as a window bench. The blinds were partially open and the morning sun added a phosphorescent glow to the smooth, grey floor. Stacey had spent the night on the bed, while Keith slept in the chair; they had traded positions after Isabella was taken for surgery. Waiting had now replaced restless sleep. They had met with the surgeons an hour ago and knew the procedure went well. The skin grafts had been taken from the top and lower back, the left side region and the flank area; her good leg was spared from grafting and thus from a full pressure garment on that side. That would be a welcome relief with the hot summer months approaching and with a garment designed as shorts, diaper changes and potty training may be easier.

The congestion in her chest was still present, although a chest x-ray showed no abnormalities or excessive fluid buildup. At this point, she did not appear to be in severe pain except the times when the staff had to move her.

Keith's mind wandered as he planned his return trip home, which would begin sometime in the evening. In order to preserve as many vacation days as possible, he would take two days each week along with his normal two days off and then work for three days.

Stacey gazed aimlessly at a nook across the room that housed a bathtub. With no doors to separate it from the rest of the room, it appeared as an oddity. This was an excellent design in allowing the sick to be scrubbed on a regular basis with as little disruption as possible. This facility spared no expense and would be a wonderful place to work if you could withstand the suffering of the young patients.

The rumble of wheels and a request from a tiny voice drew their attention to the open door of the hospital room.

"Wanna see Elmo." It was a far better greeting than Keith and Stacey had expected to hear as Isabella was wheeled back into her room.

If she were distracted by pain at all, Isabella wasn't displaying any signs; perhaps groggy, but not distressed. She was determined to return to routine life to escape all the strangers with their machines, bright lights and masked faces which were beginning to make a furry red face with an orange nose appear normal. Her bed was moved into place and her parents happily jumped to her request.

When her Elmo video ended she said, "Wanna see again." When she became thirsty she requested a drink, though she was always made to say "please" and "thank you." Keith and Stacey wanted her to be comfortable, however they didn't want her to become a demanding tyrant calling for her pipe, bowl and fiddlers three, but unlike Princess Cole—Old King Cole had not just undergone major reconstructive surgery.

As the day wore on, the pangs of pain returned. "Want Mom to rock me little bit."

"She is in pain." Stacey's face tightened. "They need to give her something stronger, this isn't working."

Keith knew the look on Stacey's face far too well.

April 6, 2005

The silence possessed its own sound like a wayward conch shell that produces the muffled rumbling of a distant train. At times it is peaceful and reclusive and at other times ominous, it reverberated as Stacey drifted in a world where anything was possible. As the intensity of her dreams elevated, sounds that were constant, though pushed back by her slumber, began tickling at her consciousness; beckoning for her return. The squeak of nurses' shoes, distant conversations and the hum of climate control units all coaxed her out of the shadows. Slowly boundless flight was replaced by reality pressing her against an overly firm mattress and a left arm that said with a definitive tingle, "Remember me? I have been buried beneath your body for the past thirty minutes."

It was 3:15 AM and though Isabella was only a few feet away from her, she was alone in the dim light that shrouded the room. She became more aware of her breathing through the rise and fall of her chest. The cool sensation of the sheets against her skin brought to mind some long forgotten commercial about cotton. Her ears now opened to the objects in the room that she hadn't realized were making soft noises when the hospital bustled with its normal fervor. As she attempted in vain to fall back to sleep, those sounds took on the echo of a leaky faucet. Her left arm certainly had no problems sleeping. She could barely move it. Now frustration became a key component in her brush with insomnia. The nurses had been in almost every hour to check on Isabella, and the child attempted each time to protest.

The distant voices grew louder by their proximity; they were coming back to prod. Two nurses appeared in the doorway. One carried an oblong plastic tray, the other two fist-sized boxes. They surrounded Isabella's beside and removed her blankets to examine the bandages covering her donor sites and grafted areas. Isabella stirred and began mumbling in a whining, pitiful voice, "Wanna rest."

Stacey sat up in bed, drawing the attention of one of the nurses who didn't seem to have a middle-of-the-night-everyone-else-is-sleeping, quiet voice. "We have to change her dressings."

Stacey cocked her head and furled her brow. "They didn't tell me about this."

"Unfortunately, the dressings need to be changed twice a day." She began to gently pull the bandages away from the raw flesh.

"No!" Isabella's voice was wet and primal. "Wanna sleep." As the air bit against her exposed wounds, she began to scream.

Stacey's face paled and her lips curled down and away. "My God," she managed, "can't we give her something for the pain."

"We gave her Tylenol with codeine. We don't want to give her anything too strong. I know it's difficult, but right now she just has to bear through it."

At times Stacey's mouth went dry. At other times it watered as though she would vomit at any moment, not from the sight of the wounds; from her vantage point she was unable to see them.

It was the sounds of pure suffering from her daughter as her voice echoed in the space of the cavernous room. Isabella screamed, cried—pleaded. Stacey balled her blankets in her fists, clutching them through the pain she was feeling; a slow, all-devouring pain that entered her womb and pierced its way out to every nerve in her body. She would never be certain if her next thought found her vocal chords, somehow escaping from between clenched teeth or if it remained shouted within the confines of her mind. It was the only rationalization she could manage to form into words. "This is barbaric!"

April 8, 2005

Two months; two eternal months. Had it been that long since she spent a blissful day shopping or brought that latest home décor project to life? She was no longer a nurse, no longer a woman, no longer a mother. Instead she was a sentry guarding against the calloused efforts of people doing a job. People who would finish their shift and return home to lives, husbands and healthy children far beyond the walls of this hospital, something she had only tasted in the past two months. Guilt stared her in the face, reflecting her loathing back to her with a sprinkle of shame.

She had been them not so long ago, caring for the sick as a means to earn a living. This experience would forever change her approach to her profession and make her an advocate for the families of her patients.

Stacey found herself at the stainless steel sink in Isabella's room, splashing cool water on her face. She wiped away the droplets with the palms of her hands, pressing deep in hopes of smoothing out the creases this ordeal had etched into her features. In the background the routines of the day began. There was blood to be drawn (some patients would need to be coaxed beyond their fear of needles), dressings to be changed and surgeries to be performed. She pressed her palms to her lower back and arched into a lengthy stretch.

Battle fatigue had worn her resolve to nothing more than Phyllo dough. The disagreements with the medical staff that began in Pittsburgh had now transferred to Cincinnati. Once again, Stacey found herself protecting Isabella against apathetic care; both real and presumed. The question was how much longer could she endure? How much longer would she have to?

"Want Mom to hold ya a little bit." Isabella's eyes were open but the cloud of sleep still drifted beneath heavy lids.

"Bird, mommy can't pick you up just yet. Pretty soon, though. Your wounds have to heal a little more." Stacey walked over and smoothed out the bed sheet. "We are going to the playroom today. Would you like that?" Isabella responded with a lazy nod.

But before that little piece of normalcy came a quick check of vitals and the irrigation of bandages. Arriving like their schedules were set to Greenwich Mean Time, two nurses who had visited exactly two hours ago entered the room to repeat the wetting of the bandages with Sulfamylon solution. And even though Stacey understood their tasks were far more necessary than the instructions on a shampoo bottle, rinse and repeat was what came to mind.

Keith would have described his wife's demeanor as "So Stacey." Arms folded across her chest; lips thinned like a rubber band that had been stretched to its limits and was awaiting the big release; brow furled in concentration of negative energy instead of deep thought. But he was not there to say anything to her; he wasn't there to give her that "you're wrong, we have to trust them" look. He would return later that day, but for now, she was alone on the frontline.

She watched as the nurses went about their tasks. Isabella emitted a soft whimper. Stacey's face tightened and her eyes widened slightly followed by a distinct and deliberate clearing of her throat. If this were Mutual of Omaha's Wild Kingdom, zoologist Marlin Perkins would be explaining how the mother bear would first send a warning growl to any intruder who approached her cub before she set loose a vicious attack.

Isabella shifted on the bed, a sudden pallor washed over her features as her tiny lips curled downward; fear pooled in her eyes. As one of the nurses tried to roll Isabella from her left side to her right, Isabella's muscles stiffened and she released a startled cry.

"She has . . ." Stacey's words were dampened by the watery cries of Isabella.

"I know this is uncomfortable, Sweetie. But this solution shouldn't hurt, okay." The nurse directed her tone to Isabella, but the words were uttered for Stacey's benefit. Soon that same logic would be used toward Stacey herself, only the words would lack the endearment allotted to the child.

Nothing beats a good drenching from a water flume ride at the amusement park on a hot summer day. The tossing of the artificial log as the turbulent water tows you along to the sheer drop-off where you make your final splash down. The payoff is a refreshing spray of water that is sometimes aided by water cannons, to ensure you are properly saturated. Take a few hours for the sun to dry your clothes and subject yourself again before you leave the park; however sitting in wet clothes in a not so dry or hot hospital and never allowing them to completely dry before the next saturation is far from a day in the park. And in the case of Isabella, her wet clothes were made up of waterlogged bandages that were pressing against the open wounds of a skin graft. It was placing a physical weight to the emotional weight that Isabella had been trying to build her new life around. So it was no small wonder when following her latest saturation she was less than enthusiastic about playing, especially from the confines of a hospital bed, and as it turns out; she was not alone.

With beige walls and gray floors polished to a mirrored sheen, the playroom was a collision of a hospital and a classroom provided you could overlook the billiard table; it sat in the middle of the room feeling awkward enough in either setting for Big Bird to sing, "One of these things is not like the others." If nothing else, the room was bright. It wasn't exactly Chuck E. Cheese's, but there were plenty of things to do. There were pictures on the walls framed with cheerfully colored borders; bookcases with shelves lined with board games, crafts, glues, paints and other delightful things in which toddlers could dip their fingers into in the name of creation. All of which could not completely hearten the children for whom the activities were intended. One toddler, who like Isabella was confined to his bed, was completely wrapped in bandages. Isabella eyed him with the angst of a child experiencing her first Halloween Trick or Treat. Accompanied by his mother and father who were attempting to interest him with activities, the young boy was in the room, but his mind was clouded by his suffering. Through an opening in the bandages, strange squeaking sounds and sometimes whimpers of pain were emitted by swollen, parched lips. Without the burden of sin, the child was already enduring all nine circles of Dante's Inferno, paying dearly for transgressions, many of which, he would never get the chance to commit. This happy visit to the playroom was a part of the child's therapy, God willing; he would remember none of it.

The room was of more than adequate size, yet the bulk of the beds allowed for little separation for children and parents alike. The other occupants in the room were a couple of Arab decent with a small girl of no more than a year. Her face was dramatically scarred from burns. One of her arms was amputated just above the elbow while the other hand had been removed at the wrist. Both were wrapped in gauze. Despite her condition, the child's attitude toward playing was more positive, suggesting to Stacey that she was a little further advanced in the healing process.

Isabella made eye contact with the girl while clutching a plush Elmo toy, but made no attempt to interact. On the adult side of things, the mother occasionally presented Stacey with a cordial smile, but made no attempt at conversation. Stacey returned the silent expression each time under the assumption the mother could not speak English while time stumbled awkwardly forward as the parents encouraged their children to play; no, forcing them, Stacey thought.

Throughout the course of the next hour Stacey coaxed Isabella with books to read, pictures to color and even a paint set; receiving only a faux smile from her daughter. After the other children left the playroom, Isabella relaxed a bit and settled on donning a fairy crown and wand and listening to music. The melodies were her greatest escape. And when her body couldn't dance, she danced with her head. The songs took away the boo-boos, sometimes even better than Mommy and Daddy could. She left herself fade away into the rhythm; allowing it to carry her up and away from the hospital bed. The music gave her the ability to completely erase the most painful memories of what had already happened and the fear of what was about to happen.

April 9, 2005

"You always have a sour puss on your face," the Nurse
Manager said.

Keith had two thoughts in mind after hearing the statement
directed toward Stacey; *here we go again* and *BULLSEYE!*

It wasn't as if they were moving in the wrong direction, it
simply seemed to Keith as soon as progress was being made there
was a setback. Two steps forward and one step back. At the end of
their stay in Pittsburgh, their relationship with the staff had calmed.
And now, with a chance to start over in a new place; they were
suddenly transported to familiar surroundings. Once again he and
Stacey found themselves in a meeting with the Nurse Manager,
the sharp dressed Dr. Eckley and the primary surgeon assigned
to Isabella's case the affable Dr. Kevin Helbley as they discussed
Stacey's demeanor. The staff in Pittsburgh had tolerated it a few
weeks before they called their meeting; here in Cincinnati they had
only waited a matter of days.

The nurse manager spoke like a t-ball coach; you're out, but in
this game everybody wins. "You have to be more positive for the
sake of the staff and for Isabella's sake." She smiled. "How would
it be if you came in one day wearing a clown nose?"

Perhaps she simply misspoke. Perhaps *Patch Adams* had just aired on *TBS* and was having a profound influence on her professional life. No matter what prompted the action, the equal and opposite reaction was obvious; a moment of silence followed by awkward laughter from the doctor's in attendance. The Nurse Manager's face flushed hot with embarrassment as she offered a shy smile. "Well, you know what I mean."

The discomfort of the moment and the stringent smell of disinfectant made the room seem smaller than its twelve foot dimensions. Keith was not claustrophobic or anti-social by nature—it was a sense of growing frustration with the asinine tiff between Stacey and the hospital staff which was causing him to want to flee the room. He was no angel by any means; a class-clown with a dark side occasionally prone to aggressive outbursts, but he could no longer endorse Stacey's need to impugn the staff's every move. To Keith, it had gone from necessary to counterproductive. And now the 'clown nose' comment simply brought out the absurdity of the entire situation.

Stacey would later tell Keith "wear a clown nose" was the most idiotic statement she had ever heard, but now she was too exhausted. She wanted to stay; knew she should stay, but maybe it was for the best. The staff wouldn't have to worry about her looking over their shoulders, Isabella wouldn't be so clingy with her not around and she could take the opportunity to check on some things at the house and organize her finances; everybody wins.

"If you want to be a part of Isabella's care and I know you do," the Nurse Manager regained her complexion and leaned back in her chair, "then you need to go home for awhile. See your family doctor about being treated for stress."

"When she is doing better; I will be doing better," Stacey said casting her eyes down at the floor; her shoulders slumped as though she would curl her body into a ball at any second.

"Isabella is out of the woods, but we have a long road of rehabilitation in front of us. We need to create a positive atmosphere to make it as easy as possible for her. We strongly feel it is best for everyone if you take a little break right now."

Stacey chewed on her lower lip as she considered the ultimatum. "That's fine. I'll go home for a few days."

"Ninety-eight percent of the skin grafts took," Dr. Helbley stated pleasantly, his tone as agreeable as his appearance. In fact, everything about the man was pleasant. His voice, his demeanor, even his face; receding hair, soft brown eyes and bushy mustache, was pleasant. His features stirred memories of actor Bill Macy who warmed television screens in the 1970's playing the role of husband and appliance store owner Walter Findlay in the sitcom *Maude*. And without Maude's sharp, deadpan, libertarian humor telling him "God will get you for that, Walter." he was free to be as pleasant as he pleased.

"There is a quarter-size area on her back that didn't take and another one and a half centimeters in the crease of her right leg which failed. These areas will heal without being covered with only a little scar tissue. All of her staples have been removed and her dressings will be dry from here on in."

"That's good news." Keith said.

Dr. Helbley nodded his agreement. "And we are going to remove her feeding tube. As long as she is able to maintain a proper diet and swallow her medication, we won't have to use it again."

Tears welled in Keith's eyes. "That's great news."

"It's a step in the right direction." Dr. Eckley added.

Stacey leaned in toward Keith, "I want to be here when they remove the tube."

Two steps forward . . .

April 10, 2005

. . . and one step back.

Taking medication when you are connected to the world by monitors, breathing apparatuses, IV lines and a feeding tube down your nose is easy; you're not presented with a choice—and a child lacks the understanding that this pill will take away the pain.

With her throat raw from the freshly removed feeding tube Isabella put up a valiant fight with the nurse and her Dad, but in the end—they overpowered her into taking a dose of methadone

The pill made it down. Then like the promo pitch for the latest product from the degenerate side of the Hasbro product development team—Baby Infirmary. "She chokes. She gags. She vomits." The pill came right back up.

And now they were faced with a foreboding question; had they just made the idea of eating worse for Isabella. The effects of two long months without food and water passing through your esophagus into your stomach does not make you hungry like a lumberjack. This system now becomes untouched tissue which no longer desires contact with food, and breaking that physiological and psychological barrier was not going to be easy with a toddler.

"Oh, boy," the nurse stated with apropos regret, "I'll talk to Dr. Bailey about what we can do about her medications."

"I just don't want that feeding tube replaced." Keith said. "Every step forward gets us closer to home, and I don't want to take a step a back."

"We'll figure something out. In the meantime, try to get her to eat as often as you can."

Isabella had always been a good eater; making the transition from formula, to rice cereal, to baby food and finally more adult cuisine. Now Stacey was unable to coax her to drink.

Stacey remained unaffected by the new struggle, knowing it was simply a part of the long recovery; Keith however had adopted it as his mission in life. He was prepared to market and sell the value of a good meal through any means. He would season every attempt with a mixture of creativity and psychology; it was a task in which he was well equipped. "Mmmm. These potato chips are tasty. Isabella, you should try one."

Her face was placid, but a sense of longing to play along with her Dad's game flickered in her eyes. She was weakening.

"I'm gonna eat'em all," he smiled, "then you're gonna want one and they'll be all gone. I just felt that last one slide down into my tummy," he said with a smack of his lips.

He pulled a fresh chip from his bag, this one barely a crumb, and was about to place it to his lips when he paused with a raise of his chin. He lowered the chip to her mouth and was pleasantly surprised when she accepted it. She tested the flavor in her mouth, showing no more interest than if the chip were a piece of cardboard.

Stacey brought over a piece of a banana they had tried on her earlier. "Here Isabella, let's try a tiny bite of banana."

Once again she opened her mouth like a baby feeling textures with her lips; more uncertainty than appetence. Stacey placed the banana fragment in and waited. Isabella gagged. Stacey recoiled to avoid the possible regurgitation which never came. The chip and banana had stayed in her stomach, although the gag had signaled her parents that it was enough food for the evening.

A tiny victory, Keith thought, *but an important one*.

Keith would stay in her room for the night, while Stacey prepared to leave the following afternoon to return home for what was supposed to be a much needed break. Being away from the hospital would only serve to increase her anxiety. She wouldn't be here to make certain she ate. She would no longer be able to monitor the staff during dressing changes and the soon to begin physical therapy. Home was here at Isabella's bedside. For now, Williamsport was nothing more than a house.

April 11, 2005

"All done!" These are the universal happy words for a child undergoing medical treatment. Keith remained in the room during Isabella's dressing change, trying hard to keep her mind off the pain. Although Isabella cried, she managed to hold a simple conversation with him. After the nurses had removed the old bandages, they tested her leg for range of motion as they prepared to begin her physical therapy.

She had a restful night, and with the dressing change finished her mood became more of a two-year-old. Although, Mom and Dad the playmates were unavailable due to the appearance of Mom and Dad the food pushers.

"Do you want some scrambled eggs?"

Isabella shook her head.

"Do you want ice cream?"

Another shake of her head.

"Do you want applesauce?"

That was greeted by a "You've got to be kidding me" grimace; of course, with no ill intent directed toward the apple growers of America.

"Do you want mashed potatoes?" Stacey asked.

"Isabella, want the mashed potatoes," she acknowledged.

Stacey's jaw dropped in a look of satisfied shock. "You want mashed potatoes?"

"Yes." She spoke the word with the distinguished pronunciation that was one of her trademarks.

"Okay," Keith stammered. "We will get you mashed potatoes."

Keith dashed from the room in a gallant quest to secure the whipped spuds and within ten minutes the Holy Grail of Idaho's creamiest had arrived. After a few bites she was satisfied.

The early afternoon was spent with Keith and Stacey alternating as bed-guard and court jester. At lunch, Isabella managed to eat two pieces of French fries and some hot dog.

Following lunch, Isabella sat on Stacey's lap, upright with Stacey supporting her back while Keith and Isabella played with some plastic, brightly colored Sesame Street figures. Elmo was at his mischievous best under the control of Muppet master Keith. But even the antics of a slightly demented Elmo grew tiring after awhile, so the family turned their attention to a stack of mail from well-wishers back home.

Isabella was elated by a little purse which came in one of the boxes. Keith and Stacey were deeply moved by a letter they received from Marge, a family friend. Isabella's cousin, Rebecca, and her class at the Hepburn-Lycoming Elementary School, sent a large package full of pictures, poems and jokes. Isabella laughed and read along as best she could while Keith and Stacey spoke the words from each card and letter. The connection to healthy children was palpable for her. Their class had sent colorful drawings of rainbows, trees, smiling dogs, a sun hovering above a playground and even a picture of Elmo, who was looking far less deranged while out from under Keith's control.

In the span of a day, the small family found a brief reprieve from the rigors of a prolonged hospital stay. It was the type of day that would be fondly remembered without the burden of the backdrop as it rose above the darkness. Even on more friendly ground, the day would have held onto its sublime allure. They would rest with continued uncertainties for what tomorrow would bring, but their dreams would be painted with smiling dogs under a warm and happy sun.

April 13, 2005

Wednesday morning broke with the promise of moderate temperatures accompanied by gusty winds, making the Winnie the Pooh forecast; a blustery Winds-day. On Monday after their playtime with Isabella, Stacey had left for Williamsport leaving Keith to turn his entire focus on trying to get Isabella to eat. Tuesday was a step backwards. Keith and Stacey had affectionately nicknamed her "The Bird" and now she was eating like one. Isabella had no interest in food, wanting only as she would request, "cold milk." As liquids go, milk provided her with adequate nourishment but only in the short term. Getting her to eat solid foods was proving to be far more challenging than Keith had expected. He had hoped they would not have to reinsert the feeding tube, but resolved to accept that possibility only as a minor setback if it occurred.

Keith set his sights now on the new day; the Winds-day with the hopes of making some inroads in bringing Isabella back to her former healthy self. This morning it was Nurse Erick who was charged with performing what Keith incessantly referred to as the "nasty" dressing change. Erick was Keith's height but a lanky frame made him appear taller than he actually stood. A front wave of thick auburn hair rendered a boyish flavor to his features. He was the most empathetic of all the nurses Keith had encountered at Shriner's and that connection to Isabella made him a fan favorite to both father and daughter. The dressing changes were painful no matter how skilled and gentle the performer's approach, but what set Erick apart was his ability to make Isabella feel they suffered and endured together. She was not alone in her pain.

As Erick injected a pain shot into the Precedex IV site, he explained to Keith that tomorrow's dressing change would be done without the aid of pain killers, although the Precedex IV would remain in place. The Precedex medication was used primarily by intensive care units because the sedative did not hinder the patient's respiration, but would do little to numb the affected areas. Keith closed his eyes and released a disquieted breath, centering himself before returning his attention to the dressing change.

The process was tedious, taking nearly thirty minutes to complete. Isabella relied on the distractions from her dad and Erick to help her through. Her reward came when Erick disconnected the IV unit to allow her the freedom to leave the room without cumbersome equipment. The Bird had been granted temporary freedom. Her breakfast would be a picnic served down the hall at her favorite painting.

In the foreground to the right, the glossy black steel of the wagon's handle and tires gives the candy apple red of the body a zesty sheen. These colors are cooled by the soft beige of the canvas walls in the background. Seated in the wagon, the daughter, propped up against padded woodened rails, offers the viewer the insouciance of youth while her bandages convey the harsh reality of mortal existence. Her presence is a collision of emotions from the soft eyes and tactile, blushed cheeks to the sorrowful vulnerability of her condition. The perspective of the father on the left seems to tether him to the outside world as he reaches for the child, offering a spoonful of pale yellow eggs. The mastery of the artist has allowed us ample evidence of an optimistic future for the child as revealed by the smile on the father's face. And yet it is the painting in the background that immerses you in the celestial omnificence of the moment as your eyes fixate on the swirling backdrop of a pastel painted moon.

Whether it was a book, a painting or its presence in the sky Isabella was as enamored by the moon as her sister had been. Sitting in the hallway beneath the painting Keith couldn't help but wonder if Samantha had somehow connected with Isabella; mentored her in a way those of us bound by earth no longer remember. Most toddlers have an uncanny ability to seek out and draw near other children, oftentimes choosing to socialize with them over their parents. Do they recognize each other from before they were born? Does the purity of their souls grant them a bond that we lose over time?

Nonetheless his picnic breakfast with Isabella was stirring memories that drew him closer to both. The menu consisted of an egg omelet, a cup of yogurt and a cold glass of milk and Keith attempted to make every bite and every drink a circus to the taste buds.

Good eating means no feeding tube, and no feeding tube brings her closer to home.

She had eaten about a quarter of her egg omelet, 3 spoons of yogurt and some milk when the next spoonful of egg set off a coughing fit. Her appetite instantly waned and the breakfast beneath the moon came to an abrupt ending.

For the rest of the day Keith launched the most creative and determined campaign to encourage a kid to eat since 1967 when a certain fast food restaurant introduced the world to a hamburger pushing clown. For lunch, he closed the door to her room in an attempt to disconnect them from the circumstances of their location. They settled in with the latest DVD release of "Veggie Tales" and Stacey's perfect menu selection of a hot dog, mashed potatoes and "cold" milk. Keith acted as lodestar by pretending to eat some of the jumbo hot dog.

"More hot dog." Isabella had requested; which was occasionally followed by, "More cold milk."

In the end, she managed to eat half the hot dog, one bite of mashed potatoes and five drinks of milk. Later that night a Daddy-daughter party would feature a movie, some potato chips, and two bites of cake. And don't forget the "cold" milk.

As the afternoon shadows stretched away from the western sun, Keith and Isabella paid a visit to the fifth floor of the Cincinnati Shriners Hospital for Children. A glass door opened to a roof top play area. Like the first load of linen removed from the clothes line, Isabella enjoyed breathing her first deep scents of spring drifting in on the fresh warm air. From the comfort of her wagon, she reached skyward feeling her hand pushed back with the strong gusts of wind the forecast had promised. This was a new dawn in the middle of the day; a reconciliation with an absent life. After being ravaged by its sinister side, her body was being reminded of the world's gentle touch. With each powerful breeze she laughed uncontrollably, opening her spirit up to the pleasant wonders of the world. It was Isabella's own little adventure. She was off on a hunt for a donkey's tail or guarding the honey stash against the dreaded Heffalumps and Woozles.

Such unabashed laughter can be contagious. Keith chuckled along as he watched his daughter enjoying this Winds-day as it lived up to its name. The riotous giggling caused Isabella to appear like she was bouncing inside her padded wagon. *Silly 'ole Bird.*

April 14, 2005

Seemingly to avoid any prolonged optimism, Isabella began experiencing severe pain through the waning hours of the evening. It robbed her of her sleep and her desire to swallow the prescribed liquid painkillers. Of the 2.5 milliliters offered to her, she only managed to ingest one.

Keith rubbed her forehead gently and sang to her in a soft, whispery voice. He distracted her with silly jokes and encouraged her with talk of the glorious day when she returned home.

Several times she fell asleep only to shift her body for comfort and cause the raw skin to stretch, setting off a new spike in pain. Through it all, the brave toddler soldier rarely vocalized her discomfort. If Jesus had ever displayed His power to Keith, it was now. Beneath Isabella's lamb, tender skin her heart beat with the strength of a lion. This girl had little need for the Wizard of Oz. She already possessed the heart and the courage. She had no use for ruby slippers. Her resilience would carry her home. Moreover, she was surrounded by an entourage of knowledgeable doctors and nurses all acting under the care and generosity of an organization dedicated to the health and welfare of children. A simple fellowship determined to provide healing over harvest and remedy over revenue all born from the concept of making being a Mason more fun.

As twilight changed its sullen face to morning, the moment Keith was dreading had arrived. Today would be Isabella's first dressing change without the aid of painkillers. It would be an ambush. Despite having the topic openly discussed in her presence, the child did not understand and could no more prepare for what was to come anymore than the Romans could when Hannibal's army attacked at the Battle of Trebia. The only thing she fully understood was when the dressing change began, it hurt—really hurt.

She had started the morning in a good mood. Even as Erick entered the room her smile was cheery and welcoming. As he prepared his tray of bandages and utensils she watched in innocent amusement. But when the old bandages began to be peeled off the raw flesh tears trickled from the corners of her eyes. She whimpered, but never screamed.

Erick shook his head and raised his eyes from his work. "For a two-year-old to act this calm under these circumstances is unusual to say the least."

"She's been amazingly tough through this whole experience." Keith leaned down to Isabella and whispered in her ear. "He's almost all done, Bird. After this we can have a breakfast party if you want." Lifting his head, he returned his comments to Erick. "She's resilient like her Uncle Ralph."

When the dressing change was over Isabella showed no interest in eating. Perhaps she viewed the procedure as an omen of what was to come or perhaps it simply set within her a mood of distrust.

In a few hours the staff was about to bring her to the brink of her tolerance while Isabella once again tested the endurance of Keith's heart.

Keith had experienced grave suffering before, when it seemed his perception of time became disjointed, when it appeared that the past, present and future existed in quiet simultaneousness. At first, there is a feeling of raw instinct driven clarity in which the helplessness of our frailty is exposed, just before all time and events are devoured into a single moment frozen in time, locking ones focus on a pinpoint of darkness which challenges everything we know about our physical world and everything within it that is good.

He had experienced it when Samantha was dying. He felt his daughter's entire existence in one fell swoop as he held her in his arms for the last time. It was all there was and all that would ever be. Time is not a tangible string of events; it is only crafted by an individual's perception and can only truly be measured by the emotion in the individual heart. Each beat takes us from past to present, and from present to past . . .

Keith had gazed up at the pallid grey sky and then down at the slick layer of ice covering the twisted, ascending road ahead. He had made several attempts to navigate the winter polished surface, but the vehicle moved less like a car and more like a wet seal. The off-road snow had been deep and now shined with a covering of crunchy, frozen crystals. There was only one way up, but it wouldn't be easy. Disappointing her was out of the question—in fact it never came to mind. He sat in the car for a moment only in preparation. Then with the heat blowing wonderful, warm air against the floorboards Keith had stepped from the car out into the icy chill.

. . . from past to present . . .

They looked up at Kelly with a courteous smile covering their trepidation. Keith was kneeling next to Isabella's wheelchair in the hallway of Shriner's Hospital. Kelly was a new face, a physical therapist assigned to begin Isabella's work on mobility and range of motion.

"Okay, Isabella." Kelly smiled. "We're going down to the gym for a little exercise."

. . . from present to past . . .

Smooth and unending, the road snaked its way up the slope—at the top it would turn and curl back down from where it came until the head swallowed the tail. This snake had been touched by the White Witch of Narnia as it coiled around the earth and now set frozen in its path. Walking on the glaring surface in sneakers had been like wearing ballroom shoes to an ice hockey game. Keith had managed to stay on his feet by moving in a rhythmic shuffle until the first sharp curl in the road lead him too close to the angled asphalt near the berm. His shuffle had turned into a slide and he lost his balance. He collapsed toward the surface with his hands outstretched catching his fall with only an inch separating his face from what had promised to be a lot of pain. He slid on all fours for a span before coming to a slow halt. His face had tightened and a breath hissed through clenched teeth. He had started to boil-over inside, but a runaway temper would not serve him well on such a slick surface. Gathering his composure he made his way to the two foot wall of crusted snow. With no path to aid him, the wintry layer had appeared cumbersome to traverse, but it would be far less treacherous than the roadway.

The vapor of exhalations had floated around his reddened cheeks, and those breaths were becoming faster and harder with each step as he trudged through the deep snow. His leg muscles shook in protest. His feet now shared the space of his shoes with clumps of icy snow. Once, he had almost lost a shoe entirely. Sometimes he would get lucky for a few steps and the crust cover would support his weight if he moved cautiously, but soon a step would send one leg deep into the fill and set him off balance. At those times it would take him a few moments to pull himself free to continue his ascension. He had paused at a tree to lean against it and look up to the summit. It had been too far to see his destination where she waited for him. He looked over his shoulder at the car in the distance, pausing only for a second with the thought. He couldn't fail her. He had to keep going. The road crossed in front of his path. He decided to take his chances again on the ice.

. . . from past to present . . .

Isabella's crying was tearing at Keith, but he knew he had to be strong and supportive. He brushed a tear from the corner of her eye with his thumb. "Kelly is your friend and she's gonna help us go home. So be brave and work hard." His words had little effect in calming her. Kelly wheeled Isabella down the hallway and into a set of double doors.

Keith drifted over to the wall and eased himself down in a chair trying to focus on the positive side of the physical therapy. The very fact that she had made it this far to her recovery stage was enough reason for optimism.

He bowed his head and closed his eyes allowing his mind to wander to other places. In the calm, a melody began to flirt with his subconscious. A series of chords formed; perhaps the conception of a new chorus. It repeated itself several times, changed slightly and then faded into the background of uncompleted work tasks followed by an almost forgotten scene from a film featuring that character actor; what's his name? Heavy set. Dark curly hair.

When Isabella screamed his eyes popped open and his head flinched in the direction of the gym.

"I want my Daddy!" her scream settled into a whimper. Then a muffled plea echoed in the hollow of the corridor. "I wanna sleep."

The hairs lifted from his neck and the acid rose in his stomach. He longed to rescue Bird from the pain she must be experiencing but he knew Kelly was doing a tough job; one that he and Stacey would never be able to accomplish on their own. Occasionally he heard the cry of "no," but it was "Daddy hold ya!" that finally drew the tears from his eyes. It was the pronoun confusion of an adorable two-year-old child sounding as though she were unsure of which desire was stronger, for him to hold her or her to hold him. She was saying it for both of them. As a tear slipped down his cheek, he fought the urge to run back to her then he fought the urge to run away. He had to stay despite the cries he would rather not hear. He had to be here for her when she reemerged.

. . . from present to past . . .

Inching his way along the glassy terrain, Keith had managed to stay upright through the next bend in the road. He had been moving north and south only making his way east, toward his ultimate objective, with each bend in the asphalt. He had begun to fall into a rhythm, moving ever more swiftly. He was far from being a threat to Apolo Ohno, but he may set a record in speed sneakering if he managed to move any faster. The one thing he hadn't counted on was no ice. When his right foot hit a bare patch of pavement the sudden deceleration was enough to set him off balance.

This time there was no chance of stopping the fall. As his feet went out from under him, he had managed to roll his body slightly. When his upper back made contact with the ground he released a quick grunt followed by a pair of four letter words that had been able to escape from his tightened jaw. Any pain he felt would be assessed after he was on his feet and confirmed that no one had witnessed his act of awkwardness.

He had stood for a moment and gathered his thoughts, coming to the grave conclusion that his sneakers would not qualify as snowshoes or ice skates. At last, he decided cutting through the snow would be more direct and less dangerous. The crest was almost in sight. He was close to her now.

. . . . from past to present . . .

Keith walked down the hallway and stared for a moment at the entry to the gym; beyond that door was his daughter. Beyond that door was suffering he may not be prepared to witness. He pushed the door open enough to gain a view of what was now transpiring. Isabella was on the floor in a crawling position, and she was visibly shaken.

Through everything Isabella had remained loyal to her Blankie, or at least to one of the Blankie twins. But now the blue, bear-blazoned Blankie, the apple of her eye, her four foot square of security in an insane world, was being held at ransom. The one article which provided her solace throughout this entire ordeal had been taken from her by the physical therapist who was offering its return only if she crawled to retrieve it. But the searing pain of movement was a high price to pay. She started to crawl, managing only a few inches before her face reddened. Tears poured forth and the cries that accompanied them were softened by her pain endured weakness. She was exhausted, frustrated to the point of giving up. In fact, had she been twenty years older she may have taken a fetal position on the ground and refused further treatment. But being a baby, she did not know the trappings of pride and freedom of choice; she relied on others to tell her what was best for her. Her head dropped in a pitiful sob as she willed her legs and arms to move despite the pain it caused.

The scene was all Keith could bear. He felt a wave of the same tearful frustration expressed on Isabella's face flush over his, he pulled back from the jamb and closed the door.

As the shrills of suffering voiced by a little girl who could not understand the importance of what was being done to her filled the hallway, all other sounds had fallen into a hush as though all the living creatures of the world held their breath in an uncomfortable quiet; cupping their hands against their ears to cower away from the audible evidence of torment. Innocence had been stripped away with the flesh and now harsh reality stretched what had replaced it. Slowly Isabella's objections decreased, her cries faded and the hum of the hospital machine replaced the assonance of her discomfort and pain. As if to officially mark the ending, the sounds of the world had returned. It had been a difficult time for Keith to endure and he shuddered with the thought of what his daughter had suffered. He rubbed his face with the palms of his hands, feeling his breathing return to normal. Isabella's wounds were still healing, the flesh raw and tender, all had been pulled to the limit of their potential; a cruel necessity to achieve the greater goal. Home.

. . . from present to past . . .

He had not been certain how long it had taken him to reach the summit, but that no longer mattered. He had made it.

The tiny girl knelt clutching her rabbit. The image forever carved on the stone marker above the name of Samantha Joyce Cole. Keith had vowed to himself that he would never miss visiting Samantha on her birthday or today; the anniversary of her passing—even if it had meant walking 1,000 miles through the snow and ice. He lowered himself to his knees and brushed away as much of the snow as possible. He ran his fingertips over the face of the stone. His cheeks had been bitten by the cold, but now the hot tears that had welled in his eyes ran warm on his face.

His voice whispered out with a trail of vapor that rose and dissipated in the chilled air. "Daddy's here, Babygirl."

. . . until they are one.

As Kelly wheeled Isabella down the hallway, Keith could see that while her face was a little red—she appeared none the worse from her therapy. Kelly greeted him with a smile, "She did great. It was a little rough at first, but it will get much easier as we go."

Keith knelt down to greet her face to face. He brushed a few hairs back from her forehead before tracing her cheek with his fingertips. She presented him with a courageous smile as if the entire session was nothing more than a mild discomfort.

He leaned forward to kiss her head and then whispered in her ear as he gave her a gentle hug. "Daddy's here."

April 16, 2005 at 07:06 PM EDT

Stacey and I could not be happier!!!

Isabella has been doing great today!

She is sitting in her chair right now next to her Mommy in my full view. Mommy is reading her a book and she is dressed in a cute little pink short outfit with her beautiful hair pulled back. Mom sure knows how to make her look like a little girl. Most of this week she looked like a wild woman with my hair styling skills and fashion sense.

She is so happy and you can just see how much she wants to do. The transformation is incredible.

Tomorrow we are going on an outing to McDonald's. Bird, Mom, Dad and a nurse.

Isabella is singing right now and we could just eat her up.

She was really good during her bath and dressing changes today. Stacey did a great job and we are really excited about the next 2 weeks. I have a feeling she will (be) walking some time in that time period. She is progressing so fast I would not be surprised if we were heading home the first week of May. Ultimately it is up to Isabella but look at all she has done. We have cried a lot of tears over these last 2 months plus. The next tears will be happy ones I just know it!!

Your prayers and good wishes have gotten us this far and we can't thank you all enough. Those people we have never met who have emailed their wishes for us we would love to see you all someday and thank you in person.

Love Keith & Stacey

April 18, 2005

Stacey was back with her daughter. She wished she had been there all along. It would have saved countless sleepless hours wondering if Isabella was comfortable; was she being treated properly. It would have saved her from being subjugated to the curiosities of others. Not that she minded the attention and concern of her friends and family, but it was as though each question was a reminder of her distance from Isabella and the denial of her natural maternal instincts to protect her child.

She had to admit, she was pleasantly surprised with Isabella's recovery. Having not been there to witness the minor day to day improvements, to her, the changes were nothing short of miraculous. Her daughter had faced down an indescribable demon and had already, at just two years of age, experienced more pain and suffering than Stacey could ever imagine having to endure. She admired her for the will to fight. Now she viewed Isabella as more than tender beauty, her daughter displayed true metal.

Growing up, a high priority had been placed on physical attractiveness; a challenge the women in Stacey's family had no difficulty in meeting. But along with that was a culture of financial prowess. And in between the code of behavior and the code of success little attention remained for spiritual growth and happiness. Happiness was found in success. God was made a part of their belief but not always a part of their lives. Hard work and diligence alone conjured positive results whether the topic was a checkbook or a relationship. God helped those who helped themselves. God creates us, but it is up to us to make ourselves. Isabella had taken a large step in these past few months in the making of herself.

With Keith back home for a few days, Stacey settled into her role as sole parent with less cynicism. Isabella was safe. The focus was now on minimizing the damage and restoring her quality of life. The surgeries had been replaced by physical therapy which was being performed a few doors down the hallway from where Stacey currently sat thumbing through a magazine with little interest other than the preoccupation of time.

Stacey looked up from the pages as Kelly, Isabella's physical therapist, came down the hall. Isabella was not with her, but Kelly wore a gratified smile; a kind of friendly "I know something you don't know."

"Would you like to see your daughter walk?" Kelly said.

Stacey rose to her feet and flipped the magazine on the empty chair, "Really!"

Stacey followed Kelly down to the PT gym, "I can't believe she is walking. I never expected her recovery to move so quick after everything she's been through." They stepped into the room and there was Isabella—seated in her chair. Grimacing in pain, but when she saw her mom she smiled through her discomfort.

"Not only did she walk, Mommy," Kelly said. "She took thirty steps on her first attempt." Kelly gently lifted Isabella from her chair and stood her on the floor. "The ligaments are so tight around her hip joint that she won't be able to walk with her normal stride. But the more she walks, the more they will loosen up and the more her walk will return to normal."

Stacey kneeled down and held out her arms. "Come here, Bird."

Kelly walked with her holding onto Isabella's hand to give her balance. Slowly, like the Tin man of Oz without his oil can, Isabella took her first two steps to Stacey. A splint placed on her right knee for support made her steps appear stiffer than they otherwise would. She cried out that it hurt, then took two more.

Two and a half months. Two and a half long months of will she be able to sit up on her own? Will her leg have to be amputated? Will her organs begin to fail? Will she even survive? And now by something Stacey could only describe as a miracle, she watched her daughter take ten steps and then felt her stumble into her embrace. The black memories had been kicked aside with each awkward step as Isabella walked out from beneath the hell she had endured into her mom's open arms.

Stacey held her as tight as she could without putting too much pressure on the wounds on her back. "I'm so proud of you."

April 23, 2005

One by one the tubes, monitors and IV lines were removed from Isabella, slowly cutting the artificial umbilical cord which had once kept her alive. And with the removal of the broviac catheter from her chest, she was reborn—free to breathe, nourish her body, walk on her own power and return home.

All the blood tests were normal; there was no explanation or diagnosis for what had caused the infection. And while they found solace in knowing Isabella had no underlying condition that triggered the attack, Keith and Stacey would remain fearful that the infection could reoccur. Prevention is never easy when you have no idea what to prevent.

For now, they focused on learning a new routine for getting their daughter ready for the day. Wound care, dressing changes, stretching, physical therapy and dressing her in pressure garments that will reduce the amount of scar tissue would be added to hair brushing and clothing and shoe selection.

For a few days, Stacey had to assist the nurses during the procedures until she became confident enough to perform the tasks herself. Once Keith returned, he was able to soothe Isabella, making it easier for Stacey to do the dressing changes and therapy.

Although being soothing was not possible when it came to putting on her pressure garments; that required strength and leverage.

The pressure garments consisted of shorts to cover the effected leg and a padded vest to protect the area of her ribs where the muscle had been cut away. The seamstress at the hospital adorned the garments with Elmo, Zowie and Burt from Sesame Street. They made her three pairs in all, each with color coordinated laces on the sleeves and pant legs. It would be a year and a half before Isabella would finally rid herself of these painfully constrictive vestments.

The pants were so tight Keith and Stacey had to hoist her into the garment and pull them up together. Each time they put the vest on they were fearful of dislocating Isabella's arm. Isabella accepted them as best she could, though she was never much of a fan of tight fitting clothes especially around the waist where the garments were as constraining as a girdle.

This morning her bath and the application of the cream on her wounds seemed as painful as the garments. But when all was finished, there she was, a precious little girl on a beautiful Saturday morning ready for her trip back home.

She made her rounds to the staff, who reminded her that goodbye was only temporary. They would see her again in two weeks when she returned for her first follow up, clinic appointment. Today they were driving home to avoid the layover in Philadelphia, but when she returned for her follow-up she would be flown by a volunteer pilot who donates his time, plane and fuel to help the children of Shriners.

Her first week back would be a busy one. There was a fundraiser the following Sunday afternoon, three days of therapy sessions in the small town of Montgomery which was a short drive from her home in Williamsport and she was scheduled to visit Geisinger Medical Center to have a suture removed.

After all the goodbyes were accomplished, she held Keith's hand and walked with him to the elevator. They told her she was going home, but she didn't seem to fully understand what the word truly meant.

Home

April 24, 2005

They started their journey at sunrise, now the sun was dropping quickly toward an ultimate extinguish over the western mountains. The minutes trickled by in mile-markers. Keith's mind could rationally predict the remaining time until they reached their destination, but his body kept asking, "Are we there yet?" with a combination of stiffness and aching. He glanced at Isabella's reflection in the rearview mirror, wondering the level of her discomfort.

As the drive progressed, Keith watched as the mountain range became more familiar. He had heard of the city of Clarion once before; Clearfield—perhaps several times in his life; Lock Haven—at least once a week and Jersey Shore was practically neighbors with his hometown of Williamsport. The miles were now falling away and Keith blinked his tired eyes, trying to stay in the moment to remain alert. He focuses on the present. He is in the car.

There is the small sign marking an exit for the tiny town of Linden where Keith's nephew and drummer for his band lives. He is anxious to jam with his band mates and return to his normal routines.

Next he sees the Harvest Moon Plaza and Restaurant, beyond you can see one of the movie screens from the Port drive-in rising above the tree line. Keith could remember being crammed in the back seat of a sedan with his siblings, watching the latest Hollywood offering. That was before the sound was broadcast over the car radios. Back then lines of silvery metal posts stood in the rolling field, each with two speakers which would be hooked inside the window of every car.

Then there is Henry's Bar-B-Que, behind the small structure the landscape slopes down as trees sprout up. Hidden from the sight of the passing cars is a baseball field where the Cole family played softball.

There is the first exit for the city of Williamsport. The green sign with the reflective white letters reads 'Fourth Street Exit ½ Mile.' Fourth Street begins or ends depending upon your prospective at the fringe of the city in a town called Newberry. If you take the exit and travel less than ten minutes east, you pass the entrance road to the Williamsport Area High School where Keith attended the final three years of his scholastic career. The entrance road winds up a mountain where the school sets cloaked by the elevation and acres of trees. Look at the school grounds; it would have better served as a prison. Of course when you think about it . . .

'Reach Road Industrial Park One Mile' if you follow the sign to that exit you drive up a slope to a T in the road. A right turn takes you back to the river road which winds its way along the banks of the Susquehanna River occasionally allowing entrance to recreational river lots. Think 'Green River' by Creedence Clearwater Revival.

There the Coles rented a river lot next to Keith's Uncle Sonny. Sonny had been a member of the US Navy. Sonny had a dock and a boat. Keith's dad discouraged his kids from even learning to swim. The Coles had a dock. But what they also had was a fire pit, a pavilion, as many ice and beverage filled coolers as they could carry, warm summer nights and plenty of family with which to enjoy them. You shouldn't drive a boat while under the influence of alcohol anyway.

A left at the T intersection takes you into the heart of Williamsport's industrial area. Back beyond the factories were small streets and homes. Back there was the house where he had lived with his first wife, Lori. Back there he had lived with his first three children. His boys Jesse and Shilo. Samantha. The swimming pool.

You now see the Mansfield exit which also leads to historic Bowman field, home of the Williamsport Bills, the Williamsport Crosscutters or whatever minor league team who happens to be contracted to that park at the time.

A little further and you reach the exit for Maynard Street which leads to the Williamsport Hospital where this whole ordeal began. To your right is a clear view of the Susquehanna River as it pushes its way under the concrete of the Maynard Street Bridge.

The next exit is for the Market Street Bridge leading into South Williamsport where the Peter J. McGovern Museum and Howard J. Lamade Stadium boasts to the world that Williamsport, Pennsylvania is home of Little League Baseball. If you continue on past Williamsport's favorite pastime, Route 15 South cuts down

Pennsylvania past Harrisburg, Gettysburg and ultimately crosses the Mason Dixon Line into Maryland and West Virginia. To the left of the exit sets the Williamsport Hampton Inn. If you could travel back in time twenty years the hotel would be replaced by Skating Plus, a roller skating rink where Keith's original band had performed on a few special occasions.

And finally there it is in the distance, the road sign reading 'Exit 25 Faxon ½ miles.'

Isabella looks forward and up at the back of her mommy's head. She looks over and up at Daddy. He is wearing glasses. He doesn't usually wear them. In between Mommy and Daddy, a ray of sunlight is coming down from the sky and is attached to the top part of the car's windshield. It is pulling them along like a string on a toy. She thinks maybe it makes the car go. She looks out of her window. A big car that's not a car is passing by, backwards. It has no windows, but lots of wheels. Now she sees the front of the car that's not a car; it has one window where the man sits. He is a big man, even bigger than daddy and he seems to like to bounce in his seat while he drives. Then it moves away behind her.

The car she is in slows down and begins a long turn. She can feel it pulling her body toward her window. She looks at her mommy's head. Some of Mommy's black hair is being blown by the wind. Some of the wind blows on Isabella's face. It is a good touch.

The Faxon exit led to a road Keith can almost maneuver while sleep-driving. At last down the alley which led to the parking area and garage behind their house. A sign in the form of a plastic green boy holding a red flag and wearing a red cap warned motorists to drive 'SLOW!' Next is their red brick garage which ends in a full view of the backyard and rear entrance to their home.

There is the white metal fence. There is Isabella's own personal playground—complete with railroad ties, pine mulch, sand box and red painted swing set. There is the house; the first level faced with a brick façade, the second level with beige siding and tan shutters. Their families had hung balloons in their yard; seventy-six in all—one for each day Isabella was in the hospital. They made it. They are now home. Keith turns to look back at Isabella and is shocked by what he sees.

Isabella looks out of the window across the seat from her. There is a garage. That's where Mommy keeps her car. Then there's a house and someone's playground. It is her playground, and it is her house. She feels like she does when Daddy returns from a trip. She sees her playground and many colorful balloons. She sees her house. Then everything looks watery.

"Aw, everything's okay, Bird. Don't cry; we're home." Keith is saying as he watches his daughter sobbing. She is not sobbing in fear or pain. It is a look Keith can only describe as exhausted relief. It is as though the hospital had become her life. This is a place that perhaps she never thought she would ever see again. Her sobbing is shaking her tired little body as the tears built of longing and suffering wash away the memories of her affliction. Isabella is overwhelmed, and she is finally home.

Isabella would now begin her lifelong battle with the aftermath of her illness; some struggles already foretold by her doctors at Shriners and other struggles yet unknown. The very fact that she had survived made Stacey feel justified for lashing out at the medical staff. Although she had experienced the difficulties of patient care herself, she knew she had to protect her daughter and would be prepared to do it again if necessary.

For Keith, Isabella's illness had changed his recent past, he would decide after an event that would take place on their return visit to Shriners that it had not changed him enough. He would try to remind himself everyday to be thankful for what he had in his life, to focus on the happy memories of what he had lost and to open his heart to the possibilities of the future. The power of positive thinking and his faith in God had carried him through the past three months; shaken, perhaps even a little fractured, but not broken.

Epilogue

By

Keith Cole

Somewhere along the border of Ohio and Pennsylvania

"Thou believest that there is one God; thou doest well: the devils also believe, and tremble.

But wilt thou know, O vain man, that faith without works is dead?"

-James 2:19-20

May 5, 2005

We were on our second day of the two day trip coming back from Cincinnati. It had been Isabella's first post release check up and it had gone well. We drove half way back to Williamsport, stayed in a hotel and then started the last four hours of the drive home. I was more optimistic than ever about Isabella's future, more hopeful than ever about Stacey and I finding some common ground in our marriage and more discouraged than ever about my growth as a human being since Isabella's illness. When I was desperate for Isabella's survival, I prayed feverishly; promising God I would be a better person. Now I had failed to live up to my promise once again.

We stopped at a general store on Interstate 80 on the Pennsylvania side of its border with Ohio. In the store, a disheveled looking old man had made me very uncomfortable as he stared at my daughter, who with my help was limping around the store looking at merchandise. As we were paying for our goods this less than presentable man approached her with a dollar, looking at her as he asked me if he could give her the money. I said "yes", even though I wanted to say "please, get away from my daughter".

After giving her the dollar he said, "I lost my little girl many years ago. Since then, every day for my daughter, I give a dollar to the first little girl I see in her memory."

The Christian would see this as a sign; perhaps an angel sent by God to test me. To the atheist or the agnostic this could be an encounter with fate but still a chance to do good for another human being. To others it could simply be karma dangling in my face waiting for me to send it back out in a positive manner or a negative one.

This lonely soul with whom I had something terrible in common was reaching out to me. I had lost a daughter; he had lost a daughter. This was a chance to help someone after I had the blessing of getting my Isabella back from certain death.

I could have bought him a cup of coffee and told him about my own misfortune fifteen years earlier and how this time things had been different. I could have told him that good things still happen and that he is not alone in missing someone he loves. I could have just asked about his daughter. What was she like? How old was she when she died? How long ago did this happen? I could have shown him the courtesy of looking him in the eye and smiling. What I did, was to take my daughter's hand and head quickly to the door. I said nothing of comfort to him. I did nothing for him. I failed the test from God. I missed my encounter with fate. I smacked this man with negative karma and it was sure to come back to me at some point in my life.

The worst part of all of this was that it didn't even dawn on me until we were miles down the road. Isabella was reading one of her books from her car seat. I was thinking about the way this man stared at my daughter when BAM, the light came on. Had I grown so little? How could I have an encounter like that at a time like this and fail to even recognize the importance of it? My heart ached for the man and for myself. I prayed for forgiveness and for a little more sensitivity the next time, should there be a next time.

I thank you stranger for the wake-up call. I hope you are fine and still giving a dollar to the first little girl you see every day.

The power of prayer is real. The power of positive thinking is real. Whatever you believe, the important thing to remember is that we need to love and serve one another while we are here. We need to try to do what is right and when we make a mistake we need to do our best to fix it. We need to be open to the experiences that surround us every day.

Thank you, Lord for all of my blessings and most of all for my second chance with Isabella.

From the time I was very young, I wanted to be a father. I'm not sure why, but it's what I wanted. Maybe I was supposed to go through this for whatever reason. I don't have a tremendous amount of confidence in my abilities, but I think I'm a pretty good Dad. I still ask my older boys now and again, just to reassure myself. My children taught me more about life and love than I could ever hope to teach them. They will define me long after I have breathed my last breath on this earth.

Thank you, Jesse and Shilo, for keeping me going after Samantha died. And thank you for being my sons. I am very proud of you both.

I believe the two distinct roles Stacey and I played during Isabella's hospitalization were not welcomed by the doctors and nurses but may have helped our daughter. The good cop, bad cop thing was not scripted but it may have been effective in reaching those with different personalities. Stacey's stern approach could have affected those who perform well under pressure. My heart on the sleeve approach may have affected those who respond well to positivity. In the end, it worked out better than we had hoped. Isabella survived, never had organ failure and did not lose her right leg. My wife and I still disagree on just about every subject you could imagine, but I still consider her to be the love of my life.

Thank you, Stacey for our wild ride together and most of all for giving me Isabella—and Sophia. Yes, we had another little girl in 2011 but that's a whole other book.

I was the 7th of 8 children and although we had little money growing up, my childhood was rich with love and experience. The person at the heart of all of this was my mother. She took what others would consider meager and made it bountiful. Thank you, mom for the love and direction and for being "Nanny" to all our children.

My brothers and sisters were and are, my best friends in life. They were 2nd fathers and mothers, protectors, confidants, baby sitters and fantastic aunts and uncles to my children. When I asked my younger brother to write this book, I admit I was taking advantage of him because I knew he wouldn't say, "No." Even if he wanted to refuse, I knew he wouldn't because, in our family, we always rally around each other. Tim knew I needed him on this and he wouldn't let me down. As he finished chapters in this book, Tim would have me read them and it was very hard to relive some of these moments. I would find myself there again, with all of the anxiety and fear pulling at my heart and my stomach. You did an amazing job, brother. Thanks to all of my siblings for your unending support in my life.

Thank you, Tim, for sticking with it over the last 6 years and delivering a beautiful tribute to our family. You are truly gifted.

After the death of Samantha, I never thought I would have another little girl. Isabella looked so much like Samantha as a toddler it was eerie. Having Isabella helped to soothe the pain of losing Samantha. No one could take the place of a lost child but it helps to have someone to whom you can give your love. When I think about death, I know I will miss my family, but I also know how happy I will be to see Samantha. When it is my time, it will bring me comfort to know I will hold her hand again, kneel down face to face and tell her how sorry her Daddy is for not being there when she fell in that pool.

The most vivid memories I have of her now, are from my dreams that came after her death. I'm trying to save her from drowning in one and in the other we are walking on a sunny day, hand in hand and I ask her, "Are you alright"? All she said was "Yes." But when I woke up with tears in my eyes I realized I had just had the perfect calming dream. Some would say it was my subconscious wrestling with the guilt I felt over her drowning. I believe that it was God letting me know that the guilt and pain were becoming too much to bear and I needed to see that Samantha was with him.

Thank you Samantha for the two short years that you spent with me, for the last kiss you asked me for and for being my little girl.

Geisenger Medical Center in Danville and Pittsburgh's Children's Hospital saved my daughter's life and for that I thank both organizations; most especially, Dr. Romanowski.

Shriner's Hospital in Cincinnati helped my daughter return to her life and home, and did it at no cost. The Shriners are a tremendous organization that exists on funding from donations. They do marvelous things with children and we could never give enough back for what they have done for our family. Isabella loves to sing and she has performed at various functions to raise money for Shriner's Hospitals. We will always be there to support the Shriner's organization and we give our deep-hearted thanks for their work in assisting the illnesses of the children.

I have changed so much through the two tragedies that defined my adulthood, the loss of my first daughter and the survival of my second. The lesson that sticks with me the most is that what you think could never happen to you, can. I wish I could say that I never take anything for granted anymore, that I live life to the fullest with my daughter Isabella at my side every moment. But that is story talk. Like everyone else, I get caught up with day to day life and have to remind myself when Isabella is getting on my last nerve, how lucky I am to have her stepping on that nerve.

Every time I pray, I thank God for Isabella's continued healing. I pray for her to have a long and happy life. I will pray that prayer until the day I die.

That is my wish for Isabella, a long and happy life.

Thank you, Isabella for being the brave, inspiring young lady you are. Thank you, Isabella, for letting me be your Daddy.

Author's Notes

My brother Keith and I grew up in a Christian family. By Christian, I mean we had a painting of Jesus hanging on our wall. We did not attend church but were left to develop our own relationship with God. Mine came from further back than I can recall. Perhaps it came from television. Perhaps it came from some sort of encryption God encoded into my DNA. Perhaps it did come from that painting of Jesus. Albeit, from where I stand today and as near as I can tell, I have always believed that Jesus Christ is the son of God.

Jesus was a hero to me and did more to shape my personality than anyone else I know. Through Him I learned pacifism is not a weakness and that love and understanding can erase ignorance.

That having been said, I wrote this book using the actual events to tell the story and in doing so, removed my spiritual bias. It was never my intention to evangelize you, the reader. My references to Jesus Christ were written in regards to Keith's beliefs or as an illustration when approaching a particular subject. If my love for Him seeped onto these pages it was strictly unintentional. However, my reasons for writing this book WERE strictly due to my love, faith and devotion to God.

I did not want to write this book. When I received the call from Keith asking me to do so, I was already one hundred pages deep in a fiction novel that I was very excited about writing. But as our conversation went on, I began to understand that I had no choice. Imagine your father giving you a gift, then asking to borrow it for a brief period. You may want to say, "No, that's mine and I am using it right now!" But that would be rude and ungrateful. Allow me to elaborate.

In the fall of 2006, I received a cell phone call from my brother Keith. He was emotionally charged with an almost impish quality in his voice. He had been listening to an audio book entitled 'The Christmas Box Miracle'; the story behind 'The Christmas Box,' written by author Richard Paul Evans.

He shut the disc off for a time trying to make sense of why he had experienced the trials of Samantha's young passing and Isabella's near brush with death. It had been almost two years since Isabella had fallen ill with 'flesh eating disease' and he was feeling compelled to share the power of prayer and positive thinking with the world.

Keith wanted to somehow pay for Isabella's healing. He wanted tell the world of his ordeal in hopes it would make a difference for other parents with sick children. He did not understand at the time that God's grace is a gift; one which cannot be repaid—only testified, which was ultimately the reason this book was written. At the time, he had been grappling with a few ideas regarding fund raisers for Shriner's Hospitals, but nothing had come to fruition. He thought about writing a book, but despite being a song writer he felt he was not quite up to the challenge.

After he exhausted his thought process, he pressed play on the CD once again. In the book 'The Christmas Box,' a cemetery's angel statue was a place of mourning for the female protagonist following the death of her son. While the book was fictional, the angel statue had once existed, but had since been lost to time. Because of the popularity of the story, a new angel statue was created and the author had chosen to dedicate the placing of the statue on a special date; December 6th, which was the date of the fictional child's death. Keith was suddenly struck; the answer was so simple he couldn't believe he had not come to the conclusion sooner. It was the date; December 6. It couldn't be just a coincidence to hear that particular date at this particular moment in time. Richard Paul Evans had dedicated the statue on the same day of the year Keith's brother had been born; and that brother was me. Knowing I had already completed one novel, Keith called me from his car to explain what had happened and to tell me he believed God had indirectly chosen me to pen Isabella's story.

Keith and I have always been each other's greatest creative supporter. As kid's, we created things together and as adults, I encourage his music and he my creative writing. So I was flattered that he would approach me to tell their story. But as I have said, I was not enthusiastic to say the least. Writing a book based on actual events is difficult and is not nearly as rewarding to me as a work of pure fiction. I wanted to cry out, "please don't do this to me." I knew I had to accept the task God now placed in front of me, because it was God who had granted me the gift of conveying my feelings on paper; a gift that had soothed my irrational, teen angst and prevented me from surrendering to a deadly sin.

As a teen, I had an insatiable appetite for romance but was left to starve by my social ineptitude. I had few friends, but blazingly—none of my acquaintances were female. Frustrated with myself and my circumstances, I made several half-hearted attempts on my life, though they were less suicidal and more designed to alert God's attention to my suffering. At the time I never realized God WAS paying attention; probably in spite of my plan to rid the world of myself. God knew what I needed long before I reached my teens and in preparation granted me the gift of self expression.

First, I began writing poetry as a means to cope with my feelings. Like most self proclaimed poets I was my only fan. But that was okay I considered poetry a gift for me. I had difficulty making friends, but I had found a friend in the form of a piece of plain white paper. I could talk to my new, silent friend about anything. It was my confidant. It was my chronicler. God had seen my plight and had blessed me with a friendship I could not find in any other human being.

I truly believe any creative writing ability I possess was given to me as a gift from God. So how could I not use that gift to praise God's name and to bear witness to His miracles.

So there I was, on the receiving end of my brother's phone call fighting myself against rejecting a story that was not blasted in the foundry of fiction because the facts were almost too bizarre to form into the steel of reality. This story was quite possibly a miracle. I had to accept the challenge and maybe the gift. Perhaps writing this story was also a gift from God, but was this the story of a miracle? And did Keith hear my birth date on the audio book as a means to propel the telling of this story out onto the masses?

In order to discern this, we should first examine your impression of the topic itself. What is a miracle to you? Are miracles nothing more than random acts of chaos that affect so many things until an unusual coalescence of circumstances appear to have been guided by a spiritual being? And if miracles are a part of reality, how can we possibly recognize them?

Isabella herself imparted two stories (Neither of which found their way into this book.) that have guided me to my own conclusions. The first story was something she had said to Keith a few weeks after being released from hospital care.

Keith had been reading a bedtime story to the then two-year-old Isabella when she made the following statement: *"God had two balloons. God kept the red one and gave Isabella the white one. Then Isabella floated up to the sky."*

The second instance was while she was at our mother's house, or as you now know her as Nanny. They had been playing in the living room when Isabella paused and began to slip behind a large self-standing, artificial fireplace. She explained to Nanny, *"Wait a minute Nanny. Isabella has to go talk to the angels."*

I will leave these simple tales as they are without drawing any further conclusions. You are free to make the decision as to the spiritual nature of Isabella's survival and to the stories told by a two-year-old girl who had experienced a brush with death.

I would like to end by extending a special thank you to Higher Hope Ministry, Thealo Rivera and Pastor Mark Gittens for their assistance in this project. And my warmest gratitude to my friend, William Rivera for his unwavering spiritual support.

As I turn my attention back to my works of fiction, I can now say I am happy for the opportunity to be a part of presenting this story to you. I am proud of the end result and proud of the loved ones in my life who assisted me in its telling. And to my brother Keith, my creative twin and favorite songsmith, I thank you for allowing me into your world and know that each word in this book was set into place with all my love.